AF348574

Oceans and Society

Oceans and Society:
Blue Planet

Edited by

Samy Djavidnia, Victoria Cheung, Michael Ott and Sophie Seeyave

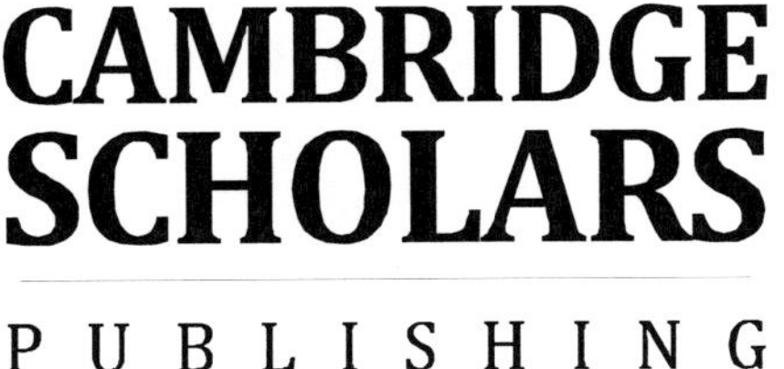

Oceans and Society: Blue Planet
Edited by Samy Djavidnia, Victoria Cheung, Michael Ott and Sophie Seeyave
Contact: Samy.Djavidnia@gmail.com

This book first published 2014

Cambridge Scholars Publishing

12 Back Chapman Street, Newcastle upon Tyne, NE6 2XX, UK

British Library Cataloguing in Publication Data
A catalogue record for this book is available from the British Library

ISBN (10): 1-4438-5639-8, ISBN (13): 978-1-4438-5639-3

TABLE OF CONTENTS

Part V: Services for the Coastal Zone

Part VI: Ocean Climate and Carbon

Part VII: Developing Capacity and Societal Awareness

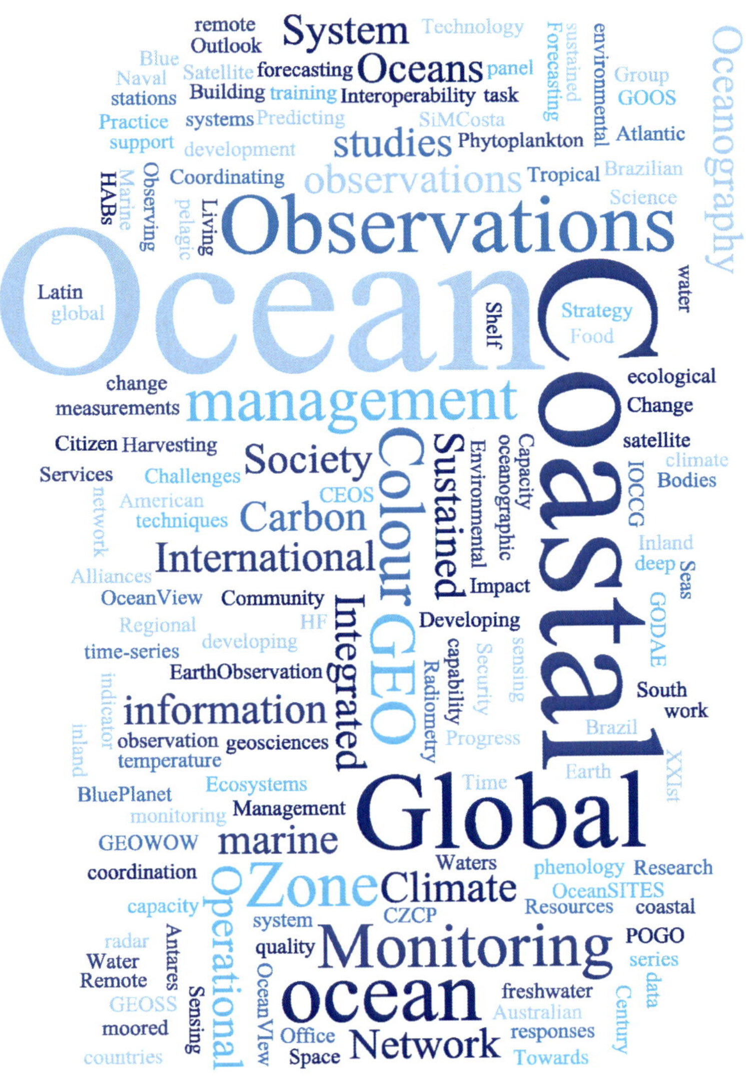

FOREWORD

BARBARA J. RYAN

"The sea, once it casts its spell, holds one in its net of wonder forever"
Jacques Cousteau

Oceans cover more than 70 percent of the Earth's surface, represent 99 percent of the planet's living space by volume and sustain life for nearly 50 percent of its entire species. The ocean works for us twenty-four hours a day, seven days a week, all year round, by producing much of the oxygen we breathe, absorbing the carbon that we create, recycling the water we drink and providing the majority of all the protein we eat. Whereas "only" 50 percent of the world population lives within 50 miles of the coast, we are all dependent on our coasts and ocean for our food, health, recreation and livelihood. Oceans and society are intricately and inextricably linked. Although humans benefit tremendously from the life-sustaining services the oceans provide, we need to increase awareness and understanding that our everyday actions impact on the ocean and its resources, and, therefore, on every one of us.

The Group on Earth Observations (GEO) is a voluntary, international, intergovernmental partnership dedicated to providing leaders in government, science, industry and civil society with accurate and timely Earth observation data and information to enable informed decision-making about the environmental challenges described above. GEO Member governments include 89 nations and the European Commission, and 67 Participating Organisations comprising international bodies with an interest or mandate in Earth observations. The GEO community is creating a Global Earth Observation System of Systems (GEOSS) that will link Earth observation resources worldwide across multiple Societal Benefit Areas, including water.

"Oceans and Society: Blue Planet" is an exciting new GEO initiative designed to:

- Raise public awareness of the role of the oceans in the Earth system, of their impacts (positive and negative) on humankind and of the societal benefits of ocean observations;
- Coordinate the various marine initiatives within GEO and develop synergies among them; and
- Advocate for and advance the establishment and maintenance of a global observing network for the oceans.

To manage our oceans and maintain ocean health and productivity, decision makers need clear, relevant, and up-to-date information. Citizens are both hungry to know, and eager to be involved actively in supporting, ocean policy and management. Scientists have the responsibility to provide the necessary building blocks for making intelligent choices and instituting good governance. We cannot do this important work without numerous national and international partnerships, collaborations and networks, and this is where the GEO "Oceans and Society: Blue Planet" initiative can play a vital role.

Sound decision-making requires that scientific knowledge be shared with society and, ultimately, with the various national and international political institutions. This book aims to provide this broader audience with information on the relevant elements of ocean observations and scientific research, as well as examples of the multiple ways in which oceans benefit society. In so doing, it is my hope that this book will contribute to bridging the existing gaps between oceans and society.

Barbara J. Ryan
Director GEO Secretariat

PREFACE

TREVOR PLATT

This book is a contribution to the activities of GEO, an intergovernmental body dedicated to developing the societal benefits of observing the Earth, either directly (*in situ*) or through the use of remote sensing. GEO is structured around nine societal-benefit areas; within each a series of Tasks is established to accomplish the overall Work Plan.

Although the oceans play important roles in each of the nine societal-benefit areas, marine affairs were initially less prominent in GEO than they deserved to be because there was no mechanism to link all of the ocean-related activities. While "green" is the colour associated with environmental responsibility and respect for planet Earth, Earth is, in fact, a "blue" planet. Exercising knowledge-based stewardship requires up-to-date information about the ocean, as well as the land and atmosphere. Therefore, it was important that, in setting its agenda, GEO recognise the importance of oceans in the Earth system.

The Partnership for Observation of the Global Oceans (POGO) lobbied strenuously over several years to stimulate a greater prominence for oceans within GEO. In May 2011, POGO submitted to the GEO Work Plan Symposium the prospectus for a new umbrella Task that would integrate and coordinate all the marine initiatives then active in GEO, as well as a number of new ones. The plan was adopted by GEO and the new Task "Oceans and Society: Blue Planet" was born, with the following mission:

The "Oceans and Society: Blue Planet Task" of GEO seeks, through the mobilisation of expert knowledge to:

- Raise public awareness of the role of the oceans in the Earth system, of their impacts (good and bad) on humankind and of the societal benefits of ocean observations;
- Coordinate the various marine initiatives within GEO and develop synergies between them; and
- Advocate and advance the establishment and maintenance of a global observing network for the oceans.

Already at the kick-off Symposium, held in Ilhabela, Brazil in November 2012, enthusiasm for Blue Planet was very strong; the new Task was definitely responding to a need. The Symposium marked the first time that all marine facets of GEO had been represented in the same room. It was an unprecedented opportunity to develop synergies among the different elements, and those who took part really felt that they had attended a landmark event.

Participants and organisers resolved to write a book showing the scope and applications of observing the ocean. This book, arising from the meeting, goes beyond mere conference proceedings. It illustrates the breadth and vitality of Earth observation in the ocean arena. It should provide a reference point, not just for marine scientists, but also for all those concerned with operational oceanography and stewardship of marine resources. The book provides an overview of the value of Earth observation in the marine sphere, from scientific advances to societal applications. It is a rich spectrum.

The Ilhabela Symposium would not have been possible without the support of the Canadian Space Agency (CSA). CSA had been funding GEO-related international programmes (SAFARI, Societal Applications in Fisheries and Aquaculture of Remote-sensing Imagery, and ChloroGIN, Chlorophyll Globally Integrated Network) for some five years; both programmes are now important elements of Blue Planet.

As Blue Planet emerged as the most significant activity in Earth observation for marine applications, CSA readily agreed to a revised funding plan that supported the Ilhabela Symposium. I am most grateful to Yves Crevier for his help in bringing this about and I hope the book will show that it was a wise decision.

Sophie Seeyave (POGO), Keith Alverson (IOC/GOOS), Boram Lee (WMO) and Douglas Cripe (GEO) helped steer the Blue Planet proposal through the approval stages at the GEO Work Plan Symposium and I thank them all. In Canada, I am grateful to Venetia Stuart (IOCCG) and Li Zhai (BIO) for their help in planning the Symposium. Shubha Sathyendranath has contributed at every stage in the evolution of Blue Planet. Milton Kampel was a wonderful local host in Brazil.

Samy Djavidnia played a vital role in helping to structure the outcomes of the Symposium, including the editing of the book, in which he was assisted by Sophie Seeyave, Vikki Cheung and Michael Ott. Albert Fischer (IOC/GOOS) presented the results of the Symposium to the GEO Plenary meeting that followed and thus helped build credibility for the fledgling Task. I thank Barbara Ryan, the new Director of GEO, for finding the time to attend the Symposium in her demanding schedule and so contribute to building confidence in Blue Planet. It was highly gratifying that so many people willingly helped in various ways to make the Symposium an undoubted success, including as authors of the chapters in this book. I am indebted to each of them.

Trevor Platt
Blue Planet Task Leader
and POGO Executive Director

PART I

INTRODUCTION

Chapter One

The Blue Planet Initiative

Samy Djavidnia, Sophie Seeyave and Trevor Platt

Our society faces a number of crucial challenges. Climate change is one of the biggest threats; sustainable management of our diverse ecosystems to enable mitigation and adaptation to these changes is imperative. Although we are accustomed to thinking that "green" is associated with environmental responsibility and protection of our Earth, Earth is in fact a "blue" planet, covered three quarters by water. As such, we need to be aware of, and understand what is happening to, both parts of the Earth's ecosystem: water as well as land. Understanding and conservation of the marine ecosystem are such essential components of global economic growth and prosperity, that the concepts and objectives of "sustainable development" make sense only if the ocean is fully incorporated.

GEOSS is a global, coordinated, comprehensive and sustained "system of observing systems" for observing the Earth on all relevant scales. Its main objective is to provide decision support tools to a wide variety of users in nine Societal Benefit Areas (SBAs). Although GEOSS is intended to cover all aspects of Earth observation, and, in this way, introduces new capabilities for monitoring and providing data on environmental processes, the Ocean component is considered a horizontal Task which spans all nine of the SBAs: Biodiversity, Ecosystems, Climate, Weather, Water, Disasters, Health, Energy and Agriculture.

The new GEO 2012–2015 Work Plan adopts a strategic target-driven structure, based on the following three pillars: 1) Infrastructure; 2) Institutions and Development, and; 3) Information for Societal Benefits. The "Information for Societal Benefits" pillar focuses on the information, services and end-to-end systems needed to support decision making across the nine SBAs. Within this pillar, the "Oceans and Society: Blue Planet" Task implements programmatic actions aimed at:

- Providing sustained ocean observations and information to underpin the development, and assess the efficacy, of global-change adaptation measures.
- Improving the global coverage and data accuracy of coastal and open-ocean observing systems.
- Coordinating and promoting the gathering, processing and analysis of ocean observations.
- Establishing a global ocean information system by making observations and information, generated on a routine basis.
- Developing a global operational ocean-forecasting network.
- Providing advanced training in ocean observations, particularly for developing countries.
- Raising awareness of biodiversity issues in the ocean.

The Blue Planet Task is a comprehensive initiative with six main Components:

- C1: Sustained Ocean Observations;
- C2: Sustained Ecosystems and Food Security;
- C3: Ocean Forecasting;
- C4: Services for the Coastal Zone;
- C5: Climate and Carbon;
- C6: Developing Capacity and Societal Awareness.

The Task Components are led primarily by the Partnership for Observation of the Global Oceans (POGO), the Committee on Earth Observation Satellites (CEOS), the Global Ocean Observing System (GOOS), the Global Ocean Data Assimilation Experiment (GODAE) OceanView, and the Coastal Zone Community of Practice (CZCP). The first effort to bring all of these different ocean (and freshwater) observing elements together was the Kick-Off Symposium in Ilhabela, São Paulo, Brazil, which took place from 19–21 November 2012. The Symposium highlighted each of the Task Components through special sessions on their programme elements and addressed a broad range of themes. The objectives of the Symposium were to:

- Learn about the diverse on-going activities;
- Better coordinate the ocean-related Tasks within GEO;
- Speak with a common voice to GEO member nations and participating organisations at the subsequent GEO Plenary in Foz do Iguaçu, Brazil;

- Raise awareness of societal benefits of ocean observations in the broader community, targeting, in particular, policymakers, and funding agencies;
- Seek new avenues for enhancing implementation of ocean observation systems; and
- Promote capacity-building globally, particularly in developing countries.

The Symposium brought together a total of 68 participants from 24 countries, comprising leaders and representatives of various international organisations and networks, research scientists and postdoctoral and graduate students. The Symposium offered the opportunity for participants to become familiar with the full scope of the "Oceans and Society: Blue Planet" Task, to develop synergies and to plan future activities. It also helped distil a clear and strong message about the way forward from the ocean community to the 2012 GEO IX Plenary (held in Foz do Iguaçu, Brazil, immediately following the Symposium). The meeting turned out to be a landmark event in the development of marine work within GEO.

This book, based on the aforementioned GEO themes and marine-related Tasks discussed at the Symposium, summarises the proposed current and future actions needed to further develop and implement the Blue Planet agenda.

This book, as was the Symposium, is structured around the Task Components, though the number of Components has grown from four to six as the Task has continued to evolve. Chapter 2 provides the GEO and GEOSS context in which the Blue Planet is embedded. Parts II to VII each comprise a number of chapters covering various aspects of each Task Component. Neither the Symposium nor the book are an exhaustive overview of the wealth of projects that are being undertaken worldwide in ocean observations and marine ecosystem and fisheries management. Rather, they are a first attempt to bring some of these projects together under a common umbrella, with the hope of entraining others as the Task evolves.

As the Blue Planet concept gathers momentum, the need for ocean indicators to provide information to citizens and society becomes fundamentally important. A shift to a paradigm under which we are able to deliver "ocean-type services" in support of all nine SBAs of GEO is required. To accomplish this, we need to listen to, and work closely with, users across the globe. This will entail a new approach, where cooperation and coordination become increasingly important, and where ocean science meets society to address issues of relevance to citizens in all nine SBAs.

The Blue Planet has recently launched its website (http://www.oceansandsociety.org), where more information on the Task can be accessed, its products downloaded and ideas both promoted and shared.

We hope this book will provide stimulating material and attract both scientists and non-experts to the field of ocean observation and marine ecosystem management. Finally, we anticipate that the book will be a valuable resource for national and international stakeholders within the marine community, including policymakers, scientists, and operational and environmental managers.

CHAPTER TWO

GEO, GEOSS AND THE 2012–2015 WORK PLAN

DOUGLAS CRIPE

In 2002, the World Summit on Sustainable Development, held in Johannesburg, South Africa, called for the integration of global observations through strengthened cooperation and coordination among global observing systems and research programmes, to help address the challenges articulated in its Plan of Implementation. These challenges included understanding the Earth system in order to enhance human health, safety and welfare, alleviate human suffering, protect the global environment and achieve sustainable development. Likewise, the summary of outcomes of the G8 Summit held in Evian, France (2003) called for strengthened international cooperation on global observation of the environment.

Following three Earth Observation Ministerial Summits (Washington DC, 2003; Tokyo, 2004; Brussels, 2005), GEO was created in 2005 as a non-binding intergovernmental partnership, with a commitment to respond to these calls through the establishment of a comprehensive, coordinated, and sustained Global Earth Observation System of Systems (GEOSS), to deliver data and information for informed decision making. GEOSS is designed to be a distributed system of systems, building on current cooperation efforts and coordination mechanisms among existing observing and processing systems, while encouraging and accommodating new components. The development of GEOSS, guided by a 10-Year Implementation Plan, is being achieved through improving and coordinating observation systems, advancing broad and open data policies and practices, fostering increased use of Earth observation data and information and building capacity. GEOSS encompasses all areas of the world and covers *in situ*, airborne, and space-based observations. Through the inclusion of a broad range of user communities, including managers and policymakers, scientific researchers and engineers, civil society, governmental and non-governmental organisations and international bodies, GEOSS is delivering information needed for informed decision-making in nine Societal Benefit Areas: Agriculture, Biodiversity, Climate, Disasters, Ecosystems, Energy, Health, Water and Weather.

Membership in GEO is open to all Member States of the United Nations. It also welcomes, as Participating Organisations (subject to approval by GEO Member States), governing bodies of the UN Specialised Agencies and Programmes, as well as intergovernmental, international and regional organisations with a mandate in Earth observation or related activities. Today, GEO counts 90 Member States and 67 Participating Organisations from across the globe, including both developed and developing nations. The main decision-making body is the full GEO Plenary, which comprises of Members and Participating Organisations and meets annually.

In its strong advocacy for broad and open data policies, GEO has established Data Sharing Principles to which its Member States and participating organisations adhere: full and open exchange of data and data and products available with minimal time delay and at minimum cost (i.e., free of charge or cost of reproduction for research and development use). The GEO Portal provides a user interface with the GEOSS Common Infrastructure (GCI), designed to facilitate the sharing of observation data and information products. The GEONETCAST system ensures that those with limited or no internet access also can have access to an increasing proportion of these data.

The GEO Work Plan provides the agreed framework through which to engage the GEOSS 10-Year Implementation Plan. It is a living document, updated annually, that represents the compendium of activities carried out, on a voluntary basis, by the Members and Participating Organisations towards the implementation of GEOSS. Specific gaps being targeted by activities of the GEO Work Plan include the uncertainty over continuity of observations, large spatial and temporal gaps in specific data sets, limited access to data and associated benefits in the developing world, inadequate data integration and interoperability, lack of relevant processing systems to transform data into useful information, inadequate user involvement and eroding or limited technical infrastructure in many parts of the world.

The activities or "Tasks" of the Work Plan are organised into three major areas according to the key objectives of GEOSS implementation to which they contribute:

- Infrastructure: the physical cross-cutting components of an operational and useable GEOSS, including interoperable observing, modelling and dissemination systems;
- Institutions and Development: describing "GEO at work" and the community's efforts to ensure that GEOSS is sustainable, relevant and widely used, with a focus on data sharing, resource mobilisation, capacity development, user engagement and science and technology integration;
- Information for Societal Benefits: the information, tools and end-to-end systems that will be made available through GEOSS to support decision-making across the nine SBAs.

In Summary: The Global Earth Observation System of Systems (GEOSS) is a coordinating and integrating network of Earth observing and information systems, to which Members States and Participating Organisations of GEO contribute on a voluntary basis, to support informed decision making for society, including the implementation of international environmental treaty obligations.

PART II

SUSTAINED OCEAN OBSERVATIONS

CHAPTER THREE

INTERNATIONAL COORDINATION OF SATELLITE OBSERVATIONS OF THE OCEAN

KERRY ANN SAWYER

Introduction

An improved understanding of the ocean system – weather, climate, ecosystems, natural resources, bathymetry and natural and human-induced hazards – is essential to better predict, mitigate, and adapt to, the expected changes to the oceans and their impacts on society, at both local and global levels. Earth observation data and derived information provide the evidence necessary for informed decision making. These support the science that underpins strategy development for local and global environmental decision making, and for monitoring progress on all geographical scales. Producing better information on the ocean environment has become a worldwide priority. International partnerships and coordination are essential to achieving this goal, because no country can monitor the vast expanse of the oceans by itself and understanding the complexities of the oceans requires global programmes and combined expertise. There are a number of key international coordination mechanisms for space-based ocean observations, including: the GEO Global Earth Observation System of Systems (GEOSS); the Global Climate Observing System (GCOS); the International Ocean Colour Coordinating Group (IOCCG); the Group on High Resolution Sea Surface Temperature (GHRSST); and the Committee on Earth Observation Satellites (CEOS). The focus of this chapter will be on CEOS and the roles it plays in international coordination of satellite observations of the ocean.

International Cooperation

Bringing space-based sensors, ground-based data analysis systems and skilled experts together requires a well-coordinated international effort and a strong commitment from space agencies. CEOS is dedicated to international collaboration among space systems and Earth observation missions and addresses the needs of the ocean community. CEOS Agencies strive to address critical scientific questions and to develop national satellite programmes with common standards and systems that can provide data to the international community, avoiding unnecessary overlaps between the different Agencies' satellite missions. CEOS ensures technical coordination among Agencies on issues concerning the usability of Earth observation data acquired by diverse systems, including coordinated access to data, inter-calibration of multiple sensors and coordination of multi-mission blended products.

CEOS, established in 1984, is an international body uniquely capable of coordinating the broad spectrum of space-based Earth observation activities. CEOS participants include government organisations that develop and operate civil Earth observation satellites (Members) and other coordinating groups and scientific or governmental organisations that support CEOS's mission (Associates). The three primary objectives of CEOS are:

1. To optimise the benefits of space-based Earth observation through cooperation of CEOS Agencies in mission planning and in the development of compatible data products, formats, services, applications and policies.
2. To aid both CEOS Agencies and the international user community by, among other things, serving as the focal point for international coordination of space-based Earth observation activities, including GEO and entities related to global change.
3. To exchange policy and technical information to encourage complementarity and compatibility among space-based Earth observation systems currently in service or development and the data received from them, as well as address issues of common interest across the spectrum of Earth observation satellite missions.

CEOS has a three working-level mechanisms to achieve the objectives listed above: CEOS Virtual Constellations; CEOS standing Working Groups; and CEOS *ad hoc* Working Groups.

CEOS Virtual Constellations typically consist of multiple space- and ground-based systems that operate together in a coordinated manner to meet a common set of observational requirements in a well-defined thematic Earth observation domain to meet societal needs. These Virtual Constellations (VCs) demonstrate the value of collaborative partnerships in addressing key observational gaps and sustaining routine collection of critical observations. There are seven CEOS VCs, of which four include marine elements: Ocean Colour Radiometry (OCR-VC); Ocean Surface Topography (OST-VC); Ocean Surface Vector Wind (OSVW-VC); and Sea Surface Temperature (SST-VC).

CEOS Working Groups typically address topics such as calibration and validation of space-based instruments and ground-based processing, common data processing standards, capacity building and data sharing, and facilitating the implementation and exploitation of time series of Essential Climate Variables (ECVs). The joint CEOS and Coordination Group for Meteorological Satellites (CGMS) Working Group on Climate (WGClimate) plays a key role in the coordination of satellite ocean observations in monitoring the oceanic ECVs and will be discussed in more detail below.

Ad hoc Working Groups are created to undertake a particular activity in support of a short-term objective. The CEOS Carbon Task Force (CTF) was established to coordinate the response from space agencies (the *Carbon Observations from Space*) to the GEO Carbon Strategy Report. This response will address the potential of space observations to monitor pools and fluxes of carbon in the ocean, land and atmospheric domains, in the context of climate change, and will provide a basis for systematic carbon observations from space and reporting of progress towards satisfying society's need for carbon information.

CEOS is deeply committed to GEO and to implementing GEOSS. This commitment is reflected in the large scale and broad scope of CEOS Agency resources allocated on a best efforts basis annually to GEO activities, including the Blue Planet Task, the Virtual Constellations and the GEO Carbon Strategy Report.

CEOS, a Participating Organisation in GEO and the "space component of GEO", established the concept of Virtual Constellations in 2006 in order to harmonise efforts among space agencies to deploy Earth observation missions, to close emerging data gaps and to contribute to GEOSS, which includes *in situ*, remotely sensed, and space-based observations. The information below, which outlines the four Virtual Constellations, is directly quoted from the *CEOS Virtual Constellations Process Paper* (updated November 2013).

The CEOS Constellation
for Ocean Colour Radiometry (OCR-VC)

The objective of the OCR-VC is to provide a time series of calibrated aquatic radiances at key wavelengths from ocean colour satellite sensors.

Well-calibrated aquatic radiances enable the estimation of many optical, biological, biogeochemical and ecological properties of Earth's aquatic environments. Activities include on-orbit and vicarious calibration, data validation, merging of satellite and *in situ* data, product generation, as well as development and demonstration of new and improved applications for scientific and management purposes.

The CEOS Constellation
for Ocean Surface Topography (OST-VC)

The objective of the OST-VC is the implementation of a sustained, systematic capability to observe the topography of, and the significant wave height on, the surface of the global oceans ranging from basin-scale to mesoscale. The OST-VC focuses on global sea level rise, the role of the oceans in climate, and operational oceanography.

The CEOS Constellation
for Ocean Surface Vector Winds (OSVW-VC)

The objective of the OSVW-VC is the implementation of a sustained, systematic capability to observe the wind field at the surface of the oceans from basin-scale to mesoscale. It focuses on the role of ocean surface wind fields in operational oceanography and meteorology, such as in supporting improvements in operational marine warnings and forecasts through the use of ocean surface vector winds from satellite scatterometry (together with significant wave height, SWH, from the OST-VC). OSVW also characterises the OSVW field for use in climate-quality data records and facilitates research related to the influence of wind forcing on the circulation of the oceans.

The CEOS Constellation
for Sea Surface Temperature (SST-VC)

The objective of the SST-VC is the development and improvement of SST products, including the SST Essential Climate Variable. SST-VC seeks to develop and implement metrics for SST services, products and users, to improve calibration and validation of the relevant instruments, and to develop training activities for satellite SST practitioners. The SST-VC serves as the formal link between CEOS and GHRSST.

Links to Societal Benefits: Climate and Carbon

GCOS has identified 48 ECVs that must be routinely monitored to meet requirements set forth under the United Nations Framework Convention on Climate Change (UNFCCC) and to contribute to the reports of the Intergovernmental Panel on Climate Change (IPCC). The *2010 GCOS Implementation Plan (GCOS IP-10), 2011 Satellite Supplement to the GCOS Implementation Plan,* and the *2012 CEOS Response to the GCOS Implementation Plan* provide detailed information on the ECVs as well as identifies the Actions that CEOS Space Agencies are undertaking to observe and monitor the ECVs.

Table 3-1: ECVs where satellite observations can make a significant contribution to coordinated outputs for monitoring of ECVs

Domain	Essential Climate Variables
Atmospheric (over land, sea, and ice)	Surface wind speed and direction; Precipitation; Upper-air temperature; Upper-air wind speed and direction; Water vapour; Cloud properties; Earth radiation budget (including solar irradiance); Carbon dioxide; Methane and other long-lived greenhouse gases; Ozone and Aerosol properties, supported by their precursors.
Oceanic	Sea-surface temperature; Sea-surface salinity; Sea level; Sea state; Sea ice; Ocean colour.
Terrestrial	Lakes; Snow cover; Glaciers and ice caps; Ice sheets; Albedo; Land cover (including vegetation type); Fraction of absorbed photosynthetically active radiation (FAPAR); Leaf area index (LAI); Above-ground biomass; Fire disturbance; Soil moisture.

The CEOS Working Group on Climate was created in 2009, and became a joint Working Group with CGMS in 2013, to coordinate and encourage collaborative activities among the CEOS Space Agencies and CGMS Meteorological Agencies in the area of climate monitoring. One of the main foci of the WGClimate is to facilitate the implementation and exploitation of ECVs through the coordination of the existing activities in the area of climate monitoring currently being undertaken by CEOS and CGMS Agencies. One of the key external stakeholders and partners of CEOS and CGMS is GCOS, with which WGClimate works very closely.

Because satellites routinely observe the entire Earth, whether circling the globe in polar orbits or focussing on a single region from geostationary orbits, they provide data and measurements to support the consistent monitoring of ocean-related ECVs. The OCR-VC, OST-VC, OSVW-VC, and the SST-VC are working very closely with the CEOS-CGMS WGClimate to develop a space-based climate information strategy, and especially to provide critical information on changes in global sea level, ocean surface vector winds, ocean colour, and SST.

Table 3-2: Overview of oceanic products that are being developed by satellite sensors

ECV	Global Product	Satellite Sensor Type
Sea Surface Temperature	Integrated SST analyses based on satellite and *in situ* data records	Infrared, Microwave
Sea Level	Sea level global mean and regional variability	Microwave
Sea State	Wave height, wind speed and direction and other measures of sea state	Altimeter
Sea Ice	Sea-ice concentration, extent, edge, supported by thickness and drift	Microwave, Visible Imager, Radar, Lidar, Synthetic Aperture Radar, Altimeter
Ocean Colour	Ocean radiometry, chlorophyll-*a*	Multispectral Visible Imager

Some of the key initiatives in ECV and climate data record (CDR) generation currently being undertaken by the four ocean-themed VCs include:

- SST-VC: generation of ECVs and CDR in collaboration with established requirements of the National Oceanic and Atmospheric Administration (NOAA) Climate Data Record Program, the European Space Agency's Climate Change Initiative (ESA CCI) Requirements, and GHRSST.
- OCR-VC: generation of ECVs and CDR in collaboration with the IOCCG.
- OST-VC: collaborating with the International Ocean Surface Topography Science Team in developing sea-level products and future mission planning.
- OSVW-VC: very operationally focused and is currently addressing the needs of the operational marine forecasting community in acquiring quality wind speed and direction data on a routine basis.

CEOS established the CTF in 2009 to coordinate the response from space agencies to the GEO Carbon Strategy Report, the *CEOS Response to the GEO Carbon Strategy*, and to:

- Take into account information requirements of both the UNFCCC and IPCC and consider how future satellite missions will support them;
- Take account of, and be consistent with, the GCOS Implementation Plans and the GEO 2012–2015 Work Plan;
- Help define the next generation carbon observing missions for individual agencies; and
- Provide a basis for systematic observation and reporting of progress towards satisfying society's need for carbon information.

The CEOS Response, which is to be released in early 2014, will address the three major domains – atmosphere, land, and oceans – and their interrelationships. Each domain chapter will include a number of Actions and Recommendations for CEOS Agencies. Some of the proposed Actions and Recommendations related to the ocean domain include: ensuring the continuity of satellite missions with adequate and sustained on-board calibration and validation operations; ensuring the high spatial resolution observations of the often turbid coastal water for analysing phytoplankton blooms; and encouraging space agencies to develop and validate products that include the ocean carbon pool.

Conclusion

CEOS is engaged in the analysis of Space Agency contributions to monitoring marine ECVs, to develop and implement space-based climate information strategies (the *CEOS Response to the GCOS IP* and the *CEOS Response to the GEO Carbon Strategy*) and especially to provide critical information on changes in global sea level, ocean surface vector winds, ocean colour, and SST, and to identify the role of the oceans in the carbon cycle and how best to define a strategy for the next generation of CEOS Agency satellite missions to meet society's needs for carbon information. CEOS is dedicated to international collaboration among space systems and Earth observation missions. It provides the multilateral coordination that enables achievement of CEOS Agency goals and addresses the needs of the ocean community.

Chapter Four

GEO High Frequency (HF) Radar

Zdenka Willis

Introduction

Just as measuring winds in the atmosphere is fundamental to weather forecasting, measuring ocean currents to determine the movement of surface waters provides critical information to support pollutant tracking, search and rescue, harmful algal bloom (HAB) monitoring, navigation and ecosystem based management and coastal and marine spatial planning. One system that has proven to effectively measure surface currents along the coast is high frequency (HF) radar.

A number of countries have used HF radar operationally for navigation, oil spill monitoring, search and rescue and HAB forecasting, but this is often done on a case-by-case basis. The United States has worked for many years to transition its HF radar network of over 130 radars to an operational system. Through its Integrated Ocean Observing System (IOOS®), the United States has succeeded in moving from individual radars, to clusters of radars to a comprehensive national network tied together through a common data architecture, a common set of practices and a national plan. Many other nations have begun to deploy HF radars and there is a tremendous amount of informal coordination and collaboration taking place. To truly make a difference on a global scale, we need to unite under a single worldwide network to make these critical measurements available for ocean and ecosystem modelling.

The United States, Australia and Spain have set forth a bold task under GEO to develop a global HF radar network. This effort was initiated during the Oceanology International Forum which took place in March 2012. At the initial meeting, a group steering the GEO HF Radar network agreed to the following long-term goals:

- To make HF radar data available in a single standardised format in near real time;

- To develop a worldwide QA/QC standard;
- To develop easy-to-use standard products;
- To assure HF radar data assimilation in ocean and ecosystem modelling;
- To develop emerging uses of HF radar in the areas of ecosystem tsunami, and climate.

Three working groups have been since established: (1) Data Management; (2) Applications and Success Stories and; (3) Best Practice in Deployment and Operation: Capacity-Building.

Several short-term activities have already been completed. United States IOOS has created an asset map to display the locations of existing HF radar sites, and has also collected the websites where HF radar data are available. See Figure 4-1. A GEO Global HF Radar webpage has been completed, http://www.ioos.noaa.gov/globalhfr/welcome.html.

A major issue confronting the HF radar community is the sharing of the already crowded frequency band. The World Radiocommunication Conference held in January–February 2012 made recommendations on frequency bands, bandwidths and their sharing (http://www.ioos.gov/hfradar/summary_wrc_12outcomes.pdf). It is now the responsibility of each country to work with their frequency managers to secure allocations. Because the HF radar signal travels a great distance, there is indeed the potential for interference. Under this effort, the co-chairs can work with the HF radar operators to provide information on the technical and regulatory aspects of frequency sharing so that radars do not interfere. This will provide the oceanographers with the information they need to enable their frequency managers to maximise operations while minimising interference.

International Cooperation

For those networks that are already established, the exchange of information on quality control, quality assurance and siting practices can be exchanged and refined. For those countries wishing to start an HF radar network we can provide these lessons learned so that they are able to establish a network more quickly. Through an existing bilateral agreement between NOAA and the government of South Korea, exchanges of personnel have taken place over the last two years and we expect that South Korea will be one of the first emerging networks to have its data portrayed on the GEO HF radar global portal.

Further, the Radiowave Operators Working Group, which is chartered to foster collaboration between new and experienced HF radar operators, to develop procedures governing HF radar operations and to provide recommendations to users, developers, manufacturers and programme managers, will be open to those involved in the GEO Task.

The following section provides an overview of the emerging HF radar networks and uses of the data.

Australia

The Australian Coastal Ocean Radar Network (ACORN) is part of a National Collaborative Research Infrastructure Strategy established by the Federal Government to enhance and streamline the funding for research infrastructure across a wide range of areas of science. ACORN, a key centre within the Integrated Marine Observing System (IMOS), operates 6 sites, a combination of WERA and CODAR SeaSonde, around the southern part of Australia.

The vessel Shen Neng 1, fully laden with 65,000 tonnes of coal and 975 tonnes of fuel oil ran aground on the Great Barrier Reef on 3 April 2010 and was re-floated on April 11. HF radar data were used to look for flotsam and pollution and to reconstruct the ocean flow and ship movement in the science report on the incident. Sick and dead fish were discovered in Gladstone Harbour within the Great Barrier Reef Marine Park. Surface current information was used to investigate the destination of (potentially toxic) harbour water as it debouched onto the shelf. Harbour water was found to extend across the shelf, but not as far as the reefs. This incident remains subject to litigation, but this demonstrates the use of high-resolution HF radar in environmental events.

China

China started the HFSWR (High-Frequency Surface Wave Radar) technology studies on sea state surveillance in the 1980s. More than 15 HFSWR stations are now operating along the Chinese coastline, including the Ocean State Monitoring and Analysing Radar series developed by Wuhan University, the CODAR SeaSonde and the Hezel WERA. Most of these stations are managed by the State Oceanic Administration and the rest by the China Meteorological Administration.

China is evaluating the data to monitor temporal and spatial variation of surface currents and as a time series to understand the seasonality of the surface current in the western part of the Taiwan Strait. When the current data from the HF radar were assimilated with other multi-sensor ocean data into the 3D numerical oceanic prediction model in the Taiwan Strait the errors between the predicted and observed data were reduced by as much as 32.4%. There have been several cases where the HF radar currents patterns have indicated the passing of typhoons; China is interested in expanding the role of HF radar in typhoon events. One of these experiments involved exploring the inversion algorithm for the sea surface's conductivity distribution by calculating the attenuation factor of radio waves along its propagation path.

Israel

Israel has installed two radars with the Hebrew University. The HF radar data are used to calculate statistics of the spatial and temporal currents as well as the dispersion of pollutants. They have used the information in real-time oil spill response drills and in basic research.

Japan

Japan began feasibility studies for oceanographic radar in the 1970s and early 1980s. The first observations of ocean waves were made in 1998. Today, 50 radars have been installed and are operated for research and monitoring of coastal areas in Japan. Ten of the sites are mobile systems for temporal observations.

The radars have been used to understand the characteristics of the Soya and Tsushima warm currents. A number of reports have been published about the variations of the Kuroshio Current. Tsunami detection is an important application of oceanographic radars that can give early warnings when the tsunami is still far offshore. Numerical experiments for tsunami detection were first introduced in 1996, but until March 2011 no-one had observed signals confirming the ability to detect tsunamis. Radars in Japan detected the tsunami at Kii Channel at a range of about 1000km from the epicentre of the Tohoku earthquake on 11 March 2011.

Korea

Korea began measuring surface currents by oceanographic radar in 1992, when CODAR SeaSonde units were used for demonstration purposes near the Keum River estuary on the west coast. Today, 25 radars operate at eight sites. The radar network near the mouth of the Keum River looks at the effects of a large coastal development on the current structure and the extension and variation of river plumes. Korea has installed radars near the Incheon port, Yeosu Bay near the major harbour of Kwangyang and in Busan New Port to support the maritime sector. Evaluation of the HF Radar data showed that the construction of the Samangeum Dyke off the western coast of Korea caused the direction of the currents to change by as much as 40 degrees. HF radar is used to observe current structure and variability such as those of the Jeju Warm Current. Korea is also using WERA systems to test its applicability in current and wave mapping off the east coast.

Malta

HF Radar was installed and made operational as part of CALYPSO, a two year project partly financed by the EU under the Operational Programme Italia–Malta 2007–2013. The main project deliverable was to establish and operate HF radar capable of recording surface currents in the Malta Channel. HF radars have been installed on the northern Malta and southern Sicilian shores and data are provided to research entities and also public entities with responsibilities for civil and environmental protection, surveillance, security and response to hazards. The data are combined with numerical models to support applications to optimise intervention in case of oil spill response, as well as to enhance tools for search and rescue, security, safer navigation, improved metro-marine forecasts, monitoring of sea conditions in critical areas such as proximity to ports and better management of the marine space between Malta and Sicily.

Spain

Spain has operated CODAR SeaSonde radars since 2009 as part of the Basque Operational data acquisition system, established by the Directorate of Meteorology and Climatology of the Basque Government. This system provides hourly surface current measurements for operational oceanography purposes in the south-eastern Bay of Biscay. In the research context, validation efforts have demonstrated the system's ability to accurately describe surface circulation at different spatial and time scales in the south-eastern Bay of Biscay. Monthly and seasonal surface circulation patterns have been described. This HF radar technology offers many benefits for the Basque Operational Oceanography System, including improvements in the knowledge of surface currents and their physical forcing processes, marine safety, search and rescue, pollution response, validation and calibration of both hydrodynamic and pollutant drift forecasting models.

The Ibiza Channel is a well-established hot spot of biodiversity related to the interaction between Atlantic and Mediterranean waters. It is a site of significant mesoscale eddy activity that affects and modifies, and occasionally blocks, the general circulation in the Western Mediterranean and associated North–South exchanges. As a result, important physical and bio-geochemical impacts are observed at different scales (from mesoscale/weeks to seasonal and inter-annual), such as Bluefin Tuna occurrence and related spawning areas, and jelly fish proliferation. The HF data, complemented with data from other ocean observing platforms, will be key to modelling the activities of these living marine resources.

Taiwan

In July 2008, the Taiwan Ocean Research Institute was established under National Applied Research Laboratories and a four year ocean technology development project was launched to map surface currents around Taiwan by using a network of HF ocean radars operating simultaneously. Taiwan has established 15 CODAR SeaSonde radar sites.

Thailand

Thailand's Geo-Informatics and Space Technology Development Agency will have 13 CODAR SeaSonde radars installed by the end of May 2013. This brings the total in Thailand to 19. They also have deployed five Wamos radars for wave field monitoring.

United States

The United States operates over 130 radars. Over 90% are CODAR SeaSondes while the remaining are WERA radars. The United States Coast Guard uses these data in their operational Search and Rescue Programme and has shown that the search area can be decreased by 66% in 96 hours, saving more lives. HF radar information was used by NOAA for oil tracking predictions during the Deepwater Horizon spill. In 2011, HF radars located in Japan and California detected and measured tsunami current flows, representing the first such tsunami observations by any radar, between 10 and 45 minutes prior to its arrival at neighbouring tide gauges.

The vessel Cosco Busan collided with the Bay Bridge in San Francisco Bay in November 2007, causing a breach in its hull which spilled fuel from its tanks. NOAA used HF radar data to determine that the spill would not reach the Farallon Islands marine sanctuary off the coast of San Francisco. This allowed spill responders to focus their equipment and manpower on more threatened areas.

In 2008, the City and County of San Francisco's wastewater system managers used HF radar; daily forecasts based on these data was used to decide whether to close nearby beaches after finding that a defect in the wastewater system was causing a point-source discharge of partially-treated wastewater. In November 2006, the City of Los Angeles used HF radar data to make decisions about beach closures during a major maintenance operation on a 50-year-old outflow pipe. City officials estimated that they could have avoided $750 million in unnecessary repairs in previous years had HF radar data been available then.

Surface current mapping has proven to be an essential tool for managers and scientists to assess and respond to HABs. HF radar is one data source used to determine the extent and severity of HABs and is among the information included in the Pacific Northwest HAB Bulletin, which provides comprehensive early warning information for Washington coast razor clam toxicity and amnesic shellfish poisoning events. Off Bodega Bay, California, researchers are using HF radar derived surface current data to obtain seasonal to annual information on ocean conditions that likely influence the survival rate of young salmon when they first enter the ocean.

Coastal surface currents can also provide important input to establishing and evaluating marine protected areas (MPAs); it is the only multi-year data with enough spatial coverage to assess how larvae of marine populations are dispersed from the location where they originate to where they settle and grow to maturity. HF radar data from a regional network in California have demonstrated the connectivity between central California MPAs by back-projecting trajectories from 10 MPAs over more than a 40-day period. Clarifying this connectivity is an important step toward understanding the movement of invertebrate and fish larvae.

The near-surface currents during the passage of Hurricane Jeanne reflected the influence of the wind as well as the Florida Current. Qualitatively, the wind-wave field as observed by the WERA HF radar responded to the rotating wind forcing as expected. It is hoped that future *in situ* measurements of wind stress and wave directional spectra in the high current shear region will provide insight as to the mechanisms associated with observed wave energy reduction.

Conclusion

GEO is fundamentally about exposing new environmental data, helping build the capacity to observe the planet and provide meaningful information to allow us to address issues of safety of life and property. The ability to provide coastal surface currents represents a huge advance in understanding coastal dynamics and can be applied to ecosystem services. One technology that has proven its merit has been the High Frequency Radar. By working internationally to build this network and provide the information in an open manner through GEO, we are able to link the networks together and leverage our collective wisdom to provide great decision-making capability in an accelerated fashion.

CHAPTER FIVE

AN OVERVIEW OF LONG-TERM OCEANOGRAPHIC MEASUREMENTS: EXISTING SITES AND EMERGING ISSUES

ALEXANDRA KRABERG
AND ANGELA SCHÄFER

Introduction

Historically, marine time series have been coastal measurements and were run by individual marine stations or laboratories. They commonly comprised just one or a small number of sites or transects and sampling frequencies ranged from work-daily to monthly (or less). However, a number of time series, such as the plankton recorder surveys of the Sir Alister Hardy Foundation for Ocean Science (SAHFOS), which commenced in the 1930s and used ships of opportunity to tow plankton recorders regularly along predefined routes, have a considerable spatial extent. The oldest plankton recorder survey covers the North Sea and North Atlantic (with regular samplings on a number of routes beginning in the North Sea in 1946) but additional surveys have also been initiated in other parts of the world, including Canada and Australia.

Time series spanning several decades exist in many parts of the world. Most of these provide information on a basic set of physico-chemical data and phyto- and/or zooplankton data, although with the advent of molecular techniques, many time series also started recording microbial population dynamics in the open ocean and deep sea. In Europe, important time-series examples include the Helgoland Roads time series, which has been sampled regularly on a work-daily basis since 1962 near the island of Helgoland in the German Bight (North Sea) (Wiltshire *et al.*, 2010). This time series provides one of the most detailed records of long-term environmental change in the German Bight. Further examples are the time series of the Zoological station Anton Dohrn and the Plymouth L4 phytoplankton time series which started in 1992 (Widdicombe *et al.*, 2010). US examples include the Narragansett and San Francisco Bay time series, while the HELCOM surveys have sampled a number of sites in all parts of the Baltic for several decades. An example in Asia includes the Seto Inland Sea time-series observatory (Takeoka 2002).

As time series are both time and labour intensive to maintain, there are many examples of time series that have been discontinued or contain gaps in their data coverage (L4, Naples). Despite their inherent limitations, these time series have already provided very valuable information on, for example, the appearance of invasive species (Boersma *et al.*, 2007) as well as HABs via long-term dynamics in local food webs, and the role of climatic drivers in determining phytoplankton dynamics (Lohmann and Wiltshire, 2012; Schlüter *et al.*, 2012). All of these time series provide a good record of long-term changes in biodiversity in relation to changes in environmental factors. As such, they are invaluable resources for assessments of the potential impacts of climate change on marine systems.

While the restriction of many time series to coastal and relatively shallow, easily accessible waters might be considered a drawback for global research, coastal areas are currently facing enormous pressure resulting from the high and ever-increasing rate of migration of human populations towards the coasts. This puts substantial pressure on ecosystem services provided by coasts and coastal waters. Valuable insights are gained from the many coastal time series around the world that can help us to adapt to and mitigate the socio-economic pressures exerted on coastal systems (Firth *et al.*, 2012; Kraberg and Wiltshire, 2013).

Until recently, individual operators usually managed these time series independently with respect to archival formats and quality control. This diversity in processing and archiving data (particularly with respect to descriptive metadata) often hampered comparative analyses of large-scale and long-term processes in related coastal areas. In the context of global networking and understanding, greater cooperation in international projects is evolving, which compels dedicated data management and integrative modules (Vandepitte *et al.*, 2010).

Emerging time series and future challenges

Time-series observations are no longer exclusively being collected and analysed manually. These data are increasingly being generated by unattended devices that are moored to the seafloor, or by floats and/or gliders (http://www.argo.ucsd.edu/). These automated systems allow us to obtain measurements from areas that are not amenable to regular observation by humans, such as the open ocean currents or the deep sea. Important examples are the OceanSITES deep moorings and the Hausgarten observatory in the Fram Strait (Soltwedel *et al.* 2005) as well as the PIES arrays in the Weddell Sea (Garzoli *et al.*, 2009; Meinen *et al.*, 2013). The time series derived from autonomous systems are shorter than the traditional, manual, time series of coastal areas, mostly not exceeding 15 years in length. Nevertheless, they already provide important information on warming trends in the deep sea (Balmaseda *et al.*, 2013) as well as the diversity of the deep-sea benthos (Soltwedel *et al.*, 2005).

These autonomous or semi-autonomous devices have not only greatly extended the spatial scale over which data are collected, but also the observation frequency, with measurements of physical parameters such as temperature or salinity commonly collected at intervals of minutes or even seconds. This development is significant for two reasons. Firstly, it facilitates a much broader range of topics that can be addressed with these data sets; secondly, data handling and archival become a much more complex and challenging task, necessitating not only the costly measuring devices themselves, but also considerable investment in workflows concerning computing, bioinformatics and data storage.

A similar challenge is the incorporation of molecular analyses into routine biodiversity time-series measurements. Molecular measurements also have very complex metadata and the interpretation of the resulting data sets depends not only on the methodologies deployed but also on the examined genes and the processing of the sequence information. Despite these difficulties, such data herald the prospect of a much more detailed understanding of biodiversity patterns in the ocean, as they facilitate the analysis of groups of organisms that are not easily discoverable by routine microscopic inspection.

The data deluge

As described above, emerging high-frequency time-series based on automated sensors require note only improved quality control efforts, but increasingly efficient data storage as well as data retrieval and re-use mechanisms. In addition, the computational difficulties involved in analysing datasets in the range of terabytes are considerable. Contrary to earlier studies, that often had gaps in data coverage, the problem is now the opposite with datasets becoming too detailed to easily extract trends or patterns of interest without additional IT or statistical skills. This problem is not restricted to genomic data but to any area of research where high frequency measurements are taken (Borgman *et al.*, 2007; Andrews,

2012). The metadata of many of these emerging time series are also very complex. Furthermore, the value of raw data is very limited; a large amount of data handling and quality assurance is often required before the data sets are usable. Nevertheless, since many of these time series are still very young, there is an opportunity for efficient harmonisation of quality control mechanisms and metadata protocols, thus avoiding some of the problems faced by older manual time series. This will require additional resources and close co-operation within and between institutions' data managers and existing repositories.

Large data repositories and digital libraries are currently seen as at least a partial solution to the problems mentioned above, in as much as they should make all data for a given topic or area discoverable and retrievable. Repositories such as ChloroGIN (http://www.chlorogin.org/, see Chapter 12), OBIS (http://www.iobis.org) and PANGAEA (http://pangaea.de) can fulfil some of these 'data needs' for certain types of data, although none can yet deal efficiently with biodiversity data. We also need to differentiate between repositories that solely archive metadata or discrete data sets and those that integrate data and provide derived data products such as maps or graphing tools summarising information from multiple data sets. Whatever their overall purpose and scope, an overarching goal of all these initiatives is greater data availability and reliability and the facilitation of broader comparisons.

International cooperation

A very recent initiative in this area is the so-called "Data Portal of German Marine Research" created by the "Marine Network of Integrated Data Access" (http://manida.org), which, as a proof of concept, aims to integrate existing German scientific and governmental marine data providers. All participating data providers are running different information technologies and data models as well as historically varying data work flows. The network does not only aim for technical integration and standardisation, but also coordinates liabilities and capacity building as well as training of its users. While this integrative portal approach will undoubtedly result in progress, the greatest challenge remains the incorporation and harmonisation of very different data types and features that make up the available information 'pool' for a given location depending on varying disciplines' specifications (see Figure 5-1). International cooperation is on the right track to develop common vocabularies and catalogues for global marine research and data management (ISO/IEC (SI), SWEET, CF, PANGAEA, NERC/BODC

(SeaDataNet), ChEBI, WORMS, ITIS, ORCID, VLIMAR, IHO, GOOS) that will help portals for cross-labelling, mapping and categorising their different historic data inventories.

The total available information 'pool' of time series continues to grow and includes not only traditional time-series information and information from automatic sensors but also molecular information, image material etc. and, in addition, historic data sets that might not necessarily satisfy modern quality control and archival standards (they often only have rudimentary metadata) (Hawkins *et al.*, 2013). For many microbial time series, dealing efficiently with all available metadata is becoming increasingly important as data analyses move more and more towards complex comparative analyses of multiple time series. It is, therefore, important that species identifications can be harmonised between time series. Geo-referenced images would provide very important metadata to support these large data integration and analysis efforts. While several databases already provide geo-referenced image information (e.g., the planktonnet database (http://planktonnet.awi.de) or the Encyclopaedia of Life (http://eol.org)) of a range of taxa, their value as regular metadata for numerical time series has not yet been recognised.

Links to societal benefits

Time-series observations from both traditional, manual time series and automated sensor systems are vital tools in the study of long-term trends in the physical and chemical environment. They provide a treasure trove of information that facilitates an understanding of past events and processes in the oceans and, based on this understanding, an assessment of the potential impacts of climate change on oceans and coasts. Therefore, time-series analyses are more than merely a matter of scientific curiosity. They are vital for managing aquatic environments and related ecosystem services such as fisheries, as well as supporting services such as tourism and recreation, not only in coastal areas but also in the open oceans.

Conclusion

The usefulness of time-series data depends on the interplay of a whole range of factors, including the use of robust quality control mechanisms throughout the process from sample collection to sample analysis and sample archival. Maintaining adequate standards in each of these areas can be a daunting task, particularly for smaller research groups. Therefore, a multidisciplinary approach to training, outreach and networking between labs/local authorities and governmental bodies is needed. Increasingly, individual institutes are providing relevant courses, but the co-ordination of such programmes to reduce fragmentation (many courses available today result from individual third-party funding as part of larger research projects) would benefit from greater involvement of international organisations such as UNESCO or GEO.

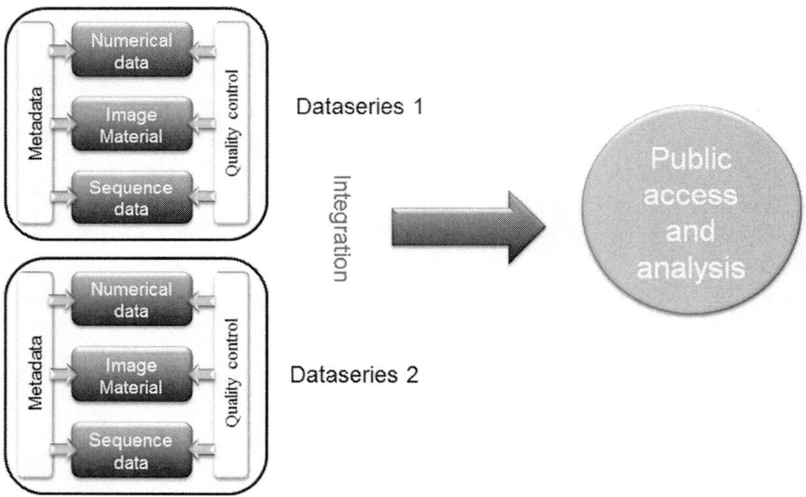

Figure 5-1: Schematics of the integrative approach applied in the Data Portal of German Marine Research

GLOBAL MONITORING OF INLAND WATER QUALITY AND FRESHWATER RESPONSES TO ENVIRONMENTAL CHANGE WITH REMOTE SENSING TECHNIQUES

TIIT KUTSER

Introduction

Current carbon-climate models, for instance those used by the IPCC or the Integrated Global Observing Strategy (IGOS), ignore inland waters, treating them as inert pipes simply transporting terrigenous organic carbon into the oceans. Recent estimates (Cole *et al.*, 2007; Tranvik *et al.*, 2009; Battin *et al.*, 2008, 2009) show that lakes are by no means inert pipes transporting carbon from land to oceans; rather, they are land carbon hot spots. Tranvik *et al.* (2009) estimated that land export of carbon to inland waters is twice as high as land export of carbon to the ocean. Most of this carbon is either subsequently exported to the oceans (0.9 Pg C y^{-1}, Figure 6-1), buried (0.6 Pg C y^{-1}), or oxidised and outgassed to the atmosphere as carbon dioxide (CO_2) and methane (CH_4) (at least 0.9 Pg C y^{-1}). Emission of methane from lakes is greater than emissions from oceans (Bastviken *et al.* 2004). Also, lake sediments may contain as much as 820 Pg C (Cole *et al.* 2007).

While estimates of carbon fluxes in inland waters remain poorly constrained compared to terrestrial and marine carbon fluxes, they are being revised at a fast pace. For instance, net CO_2 outgassing from inland waters worldwide has most recently been estimated at 3.5 Pg C y^{-1} (Aufdenkampe *et al.*, 2010). However, the implications of inland waters for the terrestrial carbon cycle (e.g., lateral fluxes) and for the marine carbon cycle (e.g., carbon sink) remain elusive. Thus, the need for integrating the inland and terrestrial carbon cycles into a 'boundless' cycle is evident and is increasingly recognised.

Ninety to ninety five percent of carbon in lakes is in the dissolved form (Wetzel, 2001). Therefore, determining the global amount of dissolved organic carbon (DOC) and the true role of lakes in the global carbon cycle requires mapping of lake DOC globally. This cannot be achieved by means of *in situ* sampling. Remote sensing is the only realistic way to obtain this information for all lakes. Only the visible part of electromagnetic radiation can penetrate the water surface and provide information regarding water properties. Consequently, only parameters that have an impact on water colour can be inferred from satellites. DOC contains coloured components called CDOM (Coloured (or *Chromophoric*) Dissolved Organic Matter). There is a strong correlation between DOC and CDOM in many water bodies (Tranvik, 1990; Kallio, 1999). Moreover, there is also a correlation between lake DOC and CO_2 supersaturation (Sobek *et al.*, 2003), meaning both DOC and the partial pressure of carbon dioxide (pCO_2) can be estimated from space using CDOM as a proxy.

Lake CDOM retrieval algorithms have been proposed (Kutser *et al.*, 2005a, b, 2009a; Kallio *et al.*, 2001, 2008). Although modelling results (Kutser *et al.*, 2005a) show good performance of the CDOM algorithms over a wide range of natural conditions, the validity of these algorithms has been tested mainly in the boreal zone. The performance of the lake CDOM-retrieval algorithms must be tested for different conditions (i.e., tropics, extremely turbid lakes, saline lakes) before global DOC estimates can be made. The launch of Landsat 8 has now made this research possible.

International Cooperation

Inland water quality studies (both *in situ* and remote sensing) are quite fragmented as water quality monitoring and inland water management are usually carried out at national and regional levels. The *in situ* community has started to consolidate its efforts through organisations like Global Lake Ecological Observatory Network (GLEON) and Networking Lake Observatories in Europe (NetLake) in order to provide input for understanding global problems and helping to solve local and regional problems in a joint effort. In many countries, both lake and river water quality measurements are part of national monitoring programmes and are funded from state budgets. This makes the effort to consolidate our knowledge about inland water quality globally slightly easier.

Inland water remote sensing is much more sophisticated than ocean remote sensing due to much higher optical complexity of the waters, more complicated atmospheric correction problems (assumptions used in ocean remote sensing are not valid) and the adjacency effect of land contaminating the water signal. Unlike the ocean remote sensing studies, that are often carried out by large institutions and in the framework of large international projects, the inland water remote sensing studies are usually carried out by small groups (often 2–3 persons) funded from national projects. Proper international cooperation between the inland water remote sensing groups is required in order to solve all the technical issues first and then start solving regional and global issues. Australia has produced a systematic analysis on the feasibility of studying inland water quality with remote sensing (Dekker and Hestir, 2012) that can be used as a basis in many countries. GEO Inland and Nearshore Coastal Water Quality Remote Sensing Group is trying to consolidate the small inland water groups around the world. However, funding such an effort is an unsolved problem and all partners have to rely on voluntary contributions by scientists around the world.

Links to societal benefits

The DOC in inland waters is important not only from a global carbon cycle point of view but also from the perspective of potential significant public health effects. DOC is a substrate for microorganisms, and hence promotes fouling of water, causing problems of taste, odour and hygiene. Moreover, disinfection of water by chlorination, which is a crucial step in maintaining water quality throughout supply systems, is quenched by DOC, resulting in the need for higher doses of chlorine. Chlorination of water rich in DOC results in the formation of carcinogenic chlorinated organic by-products (McDonald and Komulainen, 2005). In a study of more than 600,000 inhabitants in 56 Finnish towns, the Finnish Cancer Registry, together with historic information on water quality and chlorination (Koivusalo *et al.*, 1997), revealed enhanced risks of bladder and rectal cancer in connection to chlorination of water rich in DOC. A corresponding study in Norway (Magnus *et al.*, 1999), connecting the Medical Birth Registry to the Norwegian waterworks registry, demonstrated enhanced frequencies of birth defects. Accordingly, epidemiological studies imply that the increase in certain forms of cancer is a plausible side-effect of increased DOC caused by climate change. In addition, DOC has been shown to have hormone-like effects on vertebrates (Steinberg *et al.*, 2003). Thus, there is a strong need from a public health point of view to monitor changes in lake water DOC content in order to make suitable management decisions, and this cannot be done relying solely on *in situ* methods. The CDOM/DOC retrieval algorithm validated and/or developed in the framework of this project will consequently be used not only for global carbon cycle studies but also to manage drinking water resources at regional scales.

The dissolved organic carbon entering waters from surrounding land or released as a phytoplankton degradation product (up to 30% in some cases) is a food source for bacteria. DOC is also photo-oxidised in surface water due to sunlight. Carbon dioxide is released during both of these processes. Climate-carbon models (IPCC, 2007) predict warmer and wetter climate in higher latitudes. As a result, large quantities of carbon may be released as the stores of carbon locked up in northern peatlands and permafrost soils alone are equivalent to the entire pool of atmospheric CO_2. A large proportion of this carbon will be released through lakes (thaw ponds). In addition to the effects described above, CDOM also has an impact on underwater light climate: increased CDOM and detritus reduces the amount of light available for primary production as CDOM absorbs light very strongly in the blue part of the spectrum (making water brown and/or green), which is also where the absorption maximum of the main photosynthetic pigment chlorophyll-a lies. Consequently the changes in lake CDOM (due to climate change or anthropogenic forcing) may cause shifts in lake ecosystems. Higher concentrations of CDOM also cause changes in the heat budget of water as the more strongly absorbing water heats up faster.

The frequency and extent of intense phytoplankton blooms has increased in inland and coastal waters around the world (Hallegraeff 2003; Sellner *et al.*, 2003; Glibert *et al.*, 2005). Potentially harmful effects of the blooms (Edler *et al.*, 1985; Horner *et al.*, 1997; Landsberg, 2002; Backer and McGillicuddy, 2006) on human and animal health, drinking water quality and recreational use of water bodies have raised the awareness of the general public, environmental agencies and water authorities. Reliable monitoring of potentially harmful blooms is needed. Conventional monitoring networks, based on infrequent sampling in a few fixed monitoring stations, cannot provide the information needed as the blooms are very heterogeneous, both spatially and temporally. Moreover, the information is often needed over vast areas and/or large numbers of lakes. Only remote sensing can provide the spatial and temporal coverage needed. The usefulness of airborne (Wrigley and Horne, 1974) and satellite (Öström, 1976) remote sensing in detecting phytoplankton blooms was demonstrated more than three decades ago. However, how much useful information regarding blooms remote sensing can provide still remains a question. The term 'remote sensing' is used here in the broadest sense as the scale of the water bodies under investigation determines the instrumentation used. At one end of the scale are hand-held instruments, desired by water management authorities, which could rapidly detect the presence of potentially harmful blooms in small ponds and lakes. At the other end of the scale are ocean colour satellites that are the only viable option in the case of large water bodies or basin-scale studies.

In addition to operational monitoring, remote sensing could also be useful for ecosystem studies, provided quantitative remote sensing of phytoplankton blooms is feasible. Increased frequency of HABs is usually associated with eutrophication caused by anthropogenic stress (Glibert *et al.*, 2005). On the other hand, there is also the hypothesis that better wastewater treatment may increase blooms of potentially harmful cyanobacteria as better removal of nitrogen from wastewater will limit the growth of phytoplankton and will give a competitive advantage to cyanobacteria that can fix nitrogen from the atmosphere (Vahtera *et al.*, 2007). Such ecological assumptions can only be examined by time series of satellite data, provided mapping of phytoplankton biomass by remote sensing is possible. This kind of long time-series data has already been used in ecological studies (Kahru *et al.*, 2007). Use of remote sensing data also helps in understanding bloom development processes (Stumpf *et al.*, 2008).

Conclusion

At present, lake remote sensing research falls through gaps between different research fields. For example, from the remote sensing point of view optical oceanography or "ocean colour" methods have to be used to solve the problems listed above. However, the ocean colour sensors that had/have sufficient sensitivity to study water bodies (like MERIS, MODIS and future OLCI) have 300 m to 1000 m spatial resolution while the great majority of lakes are too small for such sensors. On the other hand, the land satellites that provide sufficient spatial resolution and global coverage (like previous Landsat series satellites) do not have the radiometric resolution and signal-to-noise ratio necessary for aquatic studies (Kutser, 2012). Therefore, the water quantity, not water quality, is usually addressed in the land remote sensing domain and lakes are excluded in the ocean colour research domain. The global carbon studies ignored lakes probably because they cover a relatively small area (about 3% of continental "land" surface, Downing *et al.*, 2006) and the evidence of the important role of lakes in the global carbon cycle only started to emerge after the last IPCC report was published (Tranvik *et al.*, 2009). It must be noted that the latest IPCC report (IPCC, 2013) includes inland waters in the carbon cycle with the fluxes close to the estimates provided by Tranvik *et al.* (2009). However, the correct numbers can be provided only by means of remote sensing data.

Lakes play a crucial role in humankind's everyday life and are sentinels, integrators and regulators of climate change. Remote sensing is the only feasible way to study lakes globally, and one of the key components in regional studies and water quality monitoring. These facts and new technical capabilities (suitable satellites, cheaper airborne instruments, automatic remote sensing sensors on buoys, etc.) have resulted in a significant rise in inland water remote sensing.

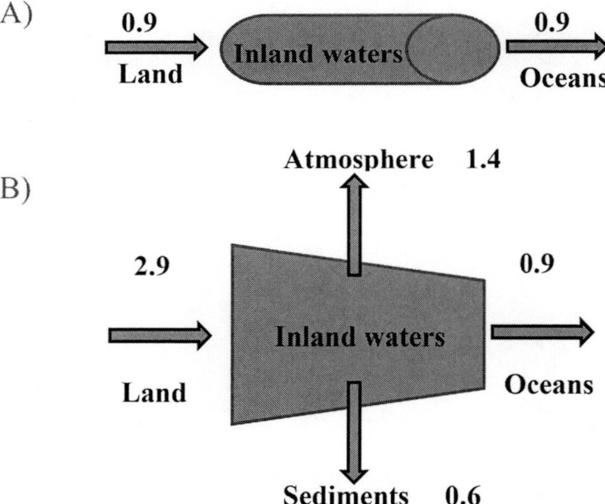

Figure 6-1: Estimated carbon fluxes through inland waters a) traditional approach used in global carbon cycle models e.g., used by the IPCC (according to Cole *et al.*, 2007); b) recent estimate (Tranvik *et al.*, 2009). Units in Pg of carbon per year

Chapter Seven

The Global Network of XBT Temperature Sections in Support of Oceanographic and Climate Studies

Gustavo Goni, Janet Sprintall,
Dean Roemmich, Ann Gronell Thresher,
Rebecca Cowley and Molly Baringer

Introduction

The eXpendable BathyThermograph (XBT) network currently provides the primary observing system for the global long-term monitoring of boundary currents. The on-going value of the XBT network is the extended time series provided by many transects as well as the unique spatially repeated sampling available along High-Density transects. The integrative relationship with other components of the ocean observing system including, for example, profiling floats, satellite altimetry, and air-sea flux measurements is also essential. Improved capabilities in ocean data assimilation and the expansion to support large-scale multidisciplinary research will further enhance the value of the XBT network into the future. Recent studies of the XBT fall rate are being evaluated with the goal of optimising the XBT historical record for climate research applications. Technological enhancements in the XBT probes are being tested to overcome fall rate issues. In addition, extensive *in situ* experiments have been carried out to reduce the depth and temperature errors of the XBT historic record. Readers are referred to the Ship Of Opportunity Program (SOOP) community white paper (Goni *et al.*, 2010), for some important scientific references that use XBT data.

Background

XBTs are widely used to observe the vertical thermal structure of the upper ocean and constitute a large fraction of the archived ocean thermal data during the 1970s, 80s, and 90s. The typical maximum sampling depth is 800m. Until the introduction of Argo profiling floats, XBTs constituted more than 50% of the global ocean thermal observations, providing sampling initially during regional research cruises and later along major shipping lines. With the Argo array now fully deployed globally, XBT observations currently represent approximately 15% of temperature profile observations. Temperature observations derived from XBTs are used to monitor the thermal structure of surface currents, in particular surface boundary currents, and are the sole practical system for monitoring the variability of mass and heat transport across fixed repeated transects.

There are three historical XBT sampling modes of deployment:

- **High Density (HD):** Four realisations per year, one XBT deployment approximately every 25km (35 XBT deployments per day with a ship speed of 20kts), aimed at obtaining high spatial resolution in one single realisation to resolve the spatial structure of mesoscale eddies, fronts, and currents.
- **Frequently Repeated (FR):** 12–18 realisations per year, six XBT deployments per day (every 100–150km), aimed at obtaining high spatial resolution observations in consecutive realisations, in regions where temporal variability is strong and resolvable with order 20-day sampling.
- **Low Density (LD):** 12 realisations per year, four XBT deployments per day (approximately every 240 km), targeted at detecting the large-scale, low frequency modes of ocean variability (now replaced by the Argo array).

There are currently a total of 55 recommended XBT transects of which 19 are in HD mode only, 23 in FR mode only, 13 in HD and FR modes, and none in LD mode (Figure 7-1). The XBT HD transects extend from ocean boundary (continental shelf) to ocean boundary, with temperature profiling at spatial separations that vary from 10 to 50 km in order to resolve fronts and boundary currents and to estimate basin-scale geostrophic velocity and mass transport integrals. PX06 (Auckland to Fiji), which began in 1986, is the earliest HD transect in the present network with almost 100 realisations, and many other HD time series now extend for more than 15 years. Detailed sampling histories and data from some of the individual open ocean HD transects are also available at: http://www-hrx.ucsd.edu, http://www.aoml.noaa.gov/phod/hdenxbt, and http://imos.org.au/underwaydata.html.

The scientific objectives of HD sampling are to:

- Measure the seasonal and inter-annual fluctuations in the transport of mass and heat, across transects which define large enclosed ocean areas and investigate their links to climate.
- Determine the long-term mean, annual cycle, inter-annual and decadal fluctuations of temperature, geostrophic velocity and large-scale ocean circulation in the top 800m of the ocean.
- Obtain long time series of temperature profiles at repeated locations in order to unambiguously separate temporal from spatial variability.
- Identify permanent boundary currents and fronts; describe their persistence and recurrence, and their relation to large-scale transports.
- Determine the space-time statistics of variability of the temperature and geostrophic shear fields.
- Estimate the significance of baroclinic eddy heat fluxes.
- Provide appropriate *in situ* data (together with Argo profiling floats, tropical moorings, air-sea flux measurements, sea level etc.) for testing ocean and ocean-atmosphere models; provide unique measurements in space and time to enhance other observing systems with unique high resolution repeat spatial sampling.

The scientific objectives of FR transects are similar to the HD objectives, however these provide greater temporary resolution and reduced spatial resolution. Hence, in addition to the objectives listed above, they are uniquely suited to:

- Initialise seasonal to inter-annual forecast models.
- Study Rossby and Kelvin wave propagation.

Scientific Achievements and Links to Societal Benefits

Societal benefits include the assessment of heat content and transport changes to better understand the extent to which the ocean sequesters heat; to identify where heat enters the ocean and where it emerges to interact with the atmosphere; and to identify changes in thermohaline circulation and monitor for indications of possible abrupt climate change. Heat content and transport influence sea-surface temperature and significant patterns of climate variability and change. Divergence of heat transports can be linked to identify changes in forcing functions driving ocean conditions and atmospheric conditions; and to elucidate oceanic influences on the global water cycle.

The XBT network provides direct estimates of the boundary current transports that could kinematically affect coastal sea level. Similarly, changes in the net meridional overturning circulation can influence long-term sea level. Extreme weather (e.g., hurricanes) and drought (e.g., rainfall) have been linked to the strength of the meridional overturning circulation. Continued analysis of boundary currents and the meridional overturning circulation and heat transport may improve forecasts of coastal inundation, seasonal forecasts of hurricanes, etc. The analysis of transbasin and closed transects provides *in situ* ocean heat and mass transport time series that can be used as a constraint on air-sea flux estimates and numerical model validation. This type of direct oceanic-transport-based flux estimate helps to reduce errors in surface products and assess numerical models.

Important research that targets the science objectives related to surface current and heat transport monitoring illustrates the unique contribution of the long-term XBT network to the global observing system. A few highlights include:

1. ***Variability of the Antarctic Circumpolar Current:*** XBT observations provide detailed information on the varying physical structure of the upper ocean across the Southern Ocean which is extremely important due to the scarcity of hydrographic observations in this region. Unique combinations of altimetry, Argo and XBT profiles have quantified the strength of the Antarctic Circumpolar Current (ACC) and shown that the Subantarctic Front contributes to over 50% of the total transport variability of the ACC, even though its net transport contribution is less than other fronts (Swart *et al.*, 2008).

2. ***Atlantic Meridional Overturning Circulation studies***: Zonal XBT lines are used to assess the meridional oceanic heat transport. For example in the centre of the subtropical gyre in the South Atlantic a clear seasonal cycle was found for the geostrophic and Ekman heat transport, which have similar amplitude but are close to $180°$ out of phase, therefore explaining the small seasonal cycle in the total northward heat transport (Garzoli *et al.*, 2013). This northward heat

transport is directly linked to the strength of the AMOC that shows a similar out of phase relationship between Ekman transport and Sverdrup transport (Dong *et al.*, 2009).

3. ***Indonesian Throughflow:*** A significant volume of Pacific Ocean water is transported through a series of narrow passages of the Indonesian seas, known as the Indonesian Throughflow (ITF), into the Indian Ocean. Data from these transects help to analyse how the ITF influences Australian climate, the Asian Monsoon, and through feedback mechanisms, impacts ENSO. ITF volumes range from 12Sv to 4Sv, depending on the strength and direction of the ENSO signal and on the Indian Ocean Dipole cycles. XBT data have also been pivotal in determining the contributions of Rossby and Kelvin waves to this phenomenon.

4. ***Temperature and geostrophic current variability in the southwest Pacific Ocean:*** XBT sections have provided 20-year time series of a closed box in the southwest Pacific that has shown the transport variability on inter-annual and decadal periods in the shallow eastward outflow from the East Australian Current, EAC (Roemmich *et al.*, 2005; Sutton *et al.*, 2001). This variability is consistent with decadal changes in wind stress that shift transport between the EAC Extension and the Tasman Front (Hill *et al.*, 2011). These data have also contributed to understanding the formation, spreading, characteristics and variability of South Pacific Subtropical Mode Water (Roemmich and Cornuelle, 1992; Tsubouchi, *et al.*, 2007).

International Cooperation

Due to the global nature of the XBT programme, the United States partners with many countries such as Australia, France, Brazil, Argentina, Italy, and South Africa in order to complete the programme objectives. The collaboration includes the sharing of resources, such as XBT probes, ship recruiting, logistics, equipment, and scientific riders. A few examples include probes provided to IRD/France that were deployed along transects in the North Atlantic Ocean (AX01 and AX20); those provided to Australia, that are used in a basin wide transect in the Indian Ocean that crosses the equator and to partly support a High Density transect between Tasmania and Antarctica; and probes provided to Italy that are deployed in important regions of the Mediterranean Sea; and equipment provided to the Federal University of Rio Grande in Brazil to carry out the AX97 section crossing the Brazil Current. The relatively complex nature of logistics, travel and changing ship routes requires substantial international agreement including perhaps most importantly data access and sharing. Most XBT profiles are automatically placed into the Global Telecommunication System (GTS) in real-time (within 24 hours). This assures that the data are available to researchers worldwide – including national weather services for use in improvements to forecast models through improved ocean state initialisation.

Ships from the SOOP provide an excellent opportunity for obtaining data from complementary observational platforms along repeated transects. Currently, Argo and surface drifters are deployed from SOOP in the XBT network. Integrating even more diverse instrumentation, the R/V Oleander is a container vessel deploying XBTs since 1976 and providing the longest temperature time series of the Gulf Stream, the Oleander also operates a Continuous Plankton Recorder (CPR) since 1975, an Acoustic Current Doppler Profiler (ADCP) that measures the upper ocean velocity since 1990, a ThermoSalinoGraph (TSG) since 1991, and a pCO_2 system since 2006. As such, this transect is now in a position to address decadal and longer variability in the characteristics and structure of this boundary current, including transport.

Future of the XBT Network

The XBT network involves the work of many components of the international field observations and science communities. The network supports the recommendations of OceanObs99 and OceanObs09, and includes several transects that the scientific community has added during the last 12 years. However, there are several challenges going forward: 1) the rapid turnover in commercial shipping and reduction in the number of ocean-crossing shipping routes; 2) the incomplete implementation of FR lines; and 3) the improvement of the XBT probe needed to reduce error in temperature and depth measurements.

A few transects are currently inactive due to implementation issues, usually related to ship recruitment, but some alternative transects are being carried out in their place. Other transects have had multi-year interruptions. Some transects may be difficult to occupy continuously due to logistical and budgetary constraints; however, they are maintained as recommendations based on the justification given by OceanObs99 and by evidence of their significant scientific contributions. Several recent transects have been initiated in the Mediterranean Sea, although their data are not currently transmitted in real-time.

The FR transects have produced noteworthy scientific insights, particularly in the eastern Indian Ocean and the Indonesian region, and represent some of the longest running time series of basin-scale ocean-structure. Nevertheless, maintenance of the many global FR transects has not been fully adopted by the scientific community. An analysis to assess the value of existing and proposed FR transects, in particular to determine the optimal sampling frequency and distance between consecutive deployments in these transects, is now being considered.

Technological advances are being planned and incorporated into the XBT operations, including the enhancement of the XBT probes to reduce temperature and depth errors and to conduct deployments in a more cost-effective fashion by improving XBT launcher systems. With the full implementation of Argo and continued altimetry observations, the role of XBTs and their impact on ocean analysis and seasonal forecasts should be re-assessed using numerical modelling and statistical analysis. Regarding real-time ocean analysis, it is critical to consider that some redundancy in the observing system is required, especially to assist automated quality control procedures. For instance, XBT data in the vicinity of Argo floats can be used as comparative data to help detect errors

Conclusion

The HD XBT network remains the primary means to obtain repeat sampling of global boundary currents, including the ACC. The network continues to increase in value through the growing number of decadal length time series, and due to the integrative relationships developed with other elements of the ocean observing system, such as:

- The implementation of global broad-scale temperature and salinity profiling by the Argo Programme, which underlines the need for complementary high density (high-resolution) data in boundary currents, frontal regions, and mesoscale eddies. HD XBT transects together with Argo provide views of the large-scale ocean interior and small-scale features near the boundary, as well as the relationship of the interior circulation to the boundary-to-boundary transport integrals.

- Almost 20 years of continuous global satellite altimetry sea-surface heights matched by contemporaneous HD sampling on many transects. The Sea Surface Height (SSH) and the subsurface temperature structure that causes most of the SSH variability are jointly measured and analysed.

- Air-sea flux estimates in large ocean areas, which complement the heat transport estimates from HD transects and the heat storage estimates from Argo.
- Improved capabilities in ocean data assimilation modelling, which allow these and other datasets to be combined and compared in a dynamically consistent framework.

Significant benefits can be achieved when complementary observations are measured from the same platform, at the same time and location. For example, integration of different observations from XBTs, TSGs, CPRs and ADCPs will be critical to understanding the variability of key ocean current systems, such as the Gulf Stream. This type of mix of instrumentation also exists on many important current systems throughout the world's oceans. This complementary sampling also provides important information as to the variability of meridional heat transport in the northern limb of the thermohaline circulation of the North Atlantic, as well as on the large changes in the sub polar gyre sink of carbon dioxide, thus providing an important component of GEOSS.

PART III

SUSTAINED ECOSYSTEMS AND FOOD SECURITY

CHAPTER EIGHT

DEVELOPING GLOBAL CAPABILITIES FOR THE OBSERVATION AND PREDICTION OF HARMFUL ALGAL BLOOMS

STEWART BERNARD, RAPHAEL KUDELA
AND LOURDES VELO-SUAREZ

Introduction

The majority of coastal regions and many freshwater systems across the globe are negatively affected by Harmful Algal Blooms (HABs). The impact of such blooms has grown in recent decades, with regard to public health, ecosystem function, fisheries and aquaculture and recreation/tourism industries (Anderson et al, 2012a). Whilst no comprehensive global study has been conducted, global economic loss due to marine HABs can be estimated at several US$ billion annually. Marine HAB-related losses in the United States are conservatively estimated at ±US$95 million annually, adjusted for inflation (Hoagland and Scatasta, 2006); analogous losses in European coastal waters are estimated at more than € 800 million (Scatasta *et al.*, 2003); HAB-related fisheries losses in Japan have been estimated at more than US$1 billion annually (Kim, 2006). In freshwater systems, potential eutrophication-related losses in the United States, primarily due to cyanobacterial blooms, are estimated at up to US$4.6 billion annually (Dodds *et al*, 2009); the 1998 season of cyanobacterial blooms in the Lake Tai catchment (China) resulted in estimated economic losses of US$6.5 billion (Le *et al.*, 2010); annual costs of freshwater algal blooms in Australia were estimated at ±A$200 million in 2000 (Atech, 2000), with similar annual eutrophication costs in the United Kingdom estimated at ±US$150 million (Pretty *et al.*, 2003) and in South Africa at ±US$250 million (Frost and Sullivan, 2010).

Although predicting the impact of global climate change on HABs is complex, the problem is likely to become more severe through range expansion of harmful species, changes in algal community dynamics and impacts in previously unaffected ecosystems (Hallegraeff, 2010) as well as by previously unknown HAB organisms (Jessup *et al.*, 2009). Increasing eutrophication through on-going development and increased resource pressure is also likely to exacerbate the global HAB problem (Heisler *et al.*, 2008), with the potential for increased proliferation of freshwater and marine cyanobacteria a major concern (O'Neil *et al.*, 2012).

Integrated Earth observation-based systems can play a significant role in the detection, monitoring and analysis of HABs in marine and freshwater ecosystems (Jochens *et al.*, 2010). The GEO Blue Planet HAB initiative seeks to consolidate and expand on existing capabilities, building a global community to develop and maximise the use and societal benefits of an integrated HAB observation and prediction system. Such a system would clearly draw strongly upon satellite-derived products, but would also seek to integrate *in situ* measurements and modelled products, disseminating integrated products through information technology systems as a component of the GEOSS architecture. Whilst the GEO HAB capability is based in Task SB-01-C2, it will offer a combination of Earth observation products, modelling forecasts and information system-based dissemination and will cut across all the Components of the Blue Planet Task, in addition to the WA-01-C4 Global Water Quality Products and Services and GEO Coastal Zone Community of Practice sub-tasks. A very important aspect of the initiative is a strong focus on community building, with the GEO Blue Planet platform seen as an essential impact mechanism for the GlobalHAB global research and development programme – the follow-up to the recently concluded 10–years Global Ecology and Oceanography of Harmful Algal Blooms (GEOHAB) programme.

The Value and Societal Benefits of a HAB Observing and Forecasting System

Global HAB-related economic losses across marine and freshwater systems can be estimated at ±US$10 billion annually. Using a typical Value Of Information (VOI) estimate of 1% of the "resource" (in this case HAB-related losses) (Macauley, 2006), a comprehensive HAB observing and forecasting information system would represent a value of ± US$100 million annually. Many of the observation system components, both observational infrastructure (space-based and *in situ*) and on-going regional projects recognising the value of these resources, are already in place – although not in a globally networked manner. The ethos of GEOSS is that improved international cooperation in the collection, interpretation and sharing of Earth observation information is an important and cost-effective mechanism for realising societal benefit (Fritz *et al.*, 2008). The Blue Planet HAB component seeks to achieve exactly this through global community- and capacity-building, leveraging both leading-edge and accessible, transferable observational and modelling approaches. The estimated VOI figure of US$100 million per annum demonstrates the considerable value of realising such a global capability. This value would increase when the value of long-term monitoring of HAB events and associated environmental drivers is included, vis-à-vis trends in HABs with potential global climate change and increasing anthropogenic pressure on coastal ecosystems.

International Cooperation

The majority of HABs occur in dynamic, rapidly-changing coastal and freshwater systems and effective HAB observation and modelling requires appreciation of a wide range of scales, trophic and allelopathic interactions and complex species-level algal physiology and behavioural characteristics. This is especially true as our awareness of HABs and our utilisation of marine resources for economic, aesthetic, and environmental reasons, expands beyond the near-shore environment (Liu *et al.*, 2009; Silver *et al.*, 2010).

The integration of the GlobalHAB/GEOHAB and the Earth observation community through Blue Planet provides an extremely valuable mechanism to quantitatively transform the research gains of the last few decades into broadly beneficial operational HAB observation systems. The principal communities involved in the activity will be GEOHAB/GlobalHAB (http://www.geohab.info), ChloroGIN and ChloroGIN-Lakes (http://www.chlorogin.org), the IOCCG (http://www.ioccg.org) and existing national projects/capabilities (e.g., NOAA CoastWatch, the US Integrated Ocean Observing System, EU initiatives such as MyOcean-linked ASIMUTH, MarCoast and AquaMar projects). There are also many active projects on the use of Earth observation for inland water quality that will make a community contribution, currently expressed through the WA-01-C4 Global Water Quality Products and Services GEO component.

An extremely important component of the initiative is capacity-building, specifically in building global community capabilities to use satellite data/products, develop low-cost *in situ* observation programmes and develop regional modelling capabilities. GEOHAB/GlobalHAB,

ChloroGIN and the IOCCG all have a proud history, existing mechanisms and strong capabilities in this regard (see Chapters 12 and 23).

Satellite Observations

Whilst ocean colour radiometry-based observations of phytoplankton biomass and other characteristics will form primary HAB observations, there is also a need for observations of physical forcing and ecosystem state variables such as SST, coastal altimetry, scatterometry-derived winds and potentially high resolution SAR-derived surface current and wind products. In terrestrially-driven freshwater systems there is a need for complementary satellite-derived observations of water cycle-related parameters such as precipitation, land cover, soil moisture and water quantity amongst others – most of which will fall under the WA-01 GEO Task. Satellite observations will provide a near real-time operational capability for bloom and ecosystem state detection, an ability to retrospectively analyse HAB occurrence from phenological and other time-series perspectives, and (ultimately) an ability to maximise model value through assimilation or integration of space-derived physical and biological data.

The current MODIS, VIIRS, OCM II and forthcoming SGLI sensors will provide a valuable fleet of ocean colour sensors for HAB observations in the GEO time frame. In Korean waters the geostationary GOCI sensor offers an extremely powerful HAB observational capability (Son *et al.*, 2012), and the high resolution hyper-spectral HICO sensor on the International Space Station offers some regionally focused demonstration capabilities (Gitelson *et al.*, 2011). Globally, the OLCI sensor on the upcoming Sentinel 3 constellation will be of particular value, allowing for frequent observation of coastal and moderate-sized inland water bodies, with 10 years of closely analogous retrospective data available from the MERIS sensor. For observation of eutrophic inland waters, the Multi Spectral Imager on the upcoming Sentinel 2 platform will offer new capabilities to derive routine water quality products at a 20m spatial resolution and < 5 day revisit time. Whilst the current ability to detect high biomass blooms will form the principal ocean-colour based capability, there is a need and potential capability to offer further description of the algal assemblage type (e.g., Brewin *et al.*, 2011) in the HAB context. There is still some debate regarding how much useful information on Phytoplankton Functional Types or Size Classes can be extracted from ocean colour data, particularly in optically complex and eutrophic waters – this is an issue that the HAB component of Blue Planet can make valuable contributions to with regard to availability of comprehensive phytoplankton assemblage and other bio-optical data (see below). Blue Planet will specifically seek to validate and make broadly available products from innovative water-type or regionally optimised algorithms.

In Situ Observations

In situ observations of HABs present particular challenges, in that there is a need for sensors that typically offer species- or toxin-specific detection. The GlobalHAB community brings considerable expertise in a variety of laboratory and autonomous-based observational techniques for assemblage, species and sub-species level, physiological, toxicological, behavioural and other aspects of phytoplankton (Anderson *et al.*, 2012a, 2012b). Two approaches will be focused on from a Blue Planet perspective, based on the GlobalHAB model that pertains to both *in situ* observations and modelling capability. The first of these are sites with leading edge autonomous capabilities to make sustained, high frequency measurements to the species or ecotoxicological level, e.g., *in situ* flow cytometers such as FlowCytoBot (Olson *et al.*, 2003), or ecogenomic sensors such as the Environmental Sample Processor (Scholin *et al.*, 2008). Such sites typically able to augment FlowCytoBot -or ESP-type measurements with both broader arrays of autonomous bio-optical sensors and sophisticated HAB physical-ecological models, will function as major satellite/model validation centres and centres of expertise for the broader community (Anderson *et al.*, 2012b).

The second approach will focus on low-cost distributable observational methods, with a particular focus on the global integration of regional *in situ* observations with ocean colour and other satellite-derived data through ChloroGIN. Many sites around the world rely on microscopic phytoplankton count data for HAB species identification – in several cases there are routinely acquired phytoplankton assemblage data across several decades (e.g., Paerl *et al.*, 2011). The availability of such data, ideally on an on-going basis and supplemented with bio-optical data, is expected to make a valuable contribution towards validation of ocean colour products in the optically-complex coastal zone. Such integration of *in situ* and satellite data is a high priority and a recognised value add for GEOSS systems (Fritz *et al.*, 2008), and a major focus area for ChloroGIN. The combination of both low-cost distributable and leading edge technology is needed to ensure global implementation of an observation network through community building, whilst providing the technological innovation necessary for sustainability.

Models and Modelling Products

One of the major value aspects of a HAB observing system is the risk-management component, and the integration of observation and modelling is critical to the ability to predict HAB occurrence (GEOHAB, 2011). While in some cases, remote sensing provides a reasonable method for detecting and tracking blooms (Stumpf and Tomlinson, 2005), there are many other cases where major ecological impact is associated with low concentrations of highly toxic species. In those circumstances satellite data and/or real-time *in situ* data offer very little traction; innovative modelling approaches offer the only prediction possibility, e.g., *Alexandrium fundyense* blooms in the Gulf of Maine (McGillicuddy *et al.*, 2011). Many HAB modelling approaches are available, ranging from simple empirical/logic-based predictions (e.g., Raine *et al.*, 2010), Neural Network and observation-based (e.g., Velo-Suárez and Gutierrez-Estrada, 2007) and a range of complex numerical models with varying degrees of data integration (e.g., Velo-Suárez *et al.*, 2010; McGillicuddy *et al.*, 2011). An analogous dual approach to the *in situ* observations (above) can be adopted: maximising the value of regions in the globe where sophisticated modelling capabilities already exist; and using this expertise to grow a global modelling community capable of using low-cost infrastructure and freely available software (e.g., Penven *et al.*, 2007). The HAB modelling component is expected to make a contribution to the larger SB-01-C3: Global Operational Ocean-Forecasting Network sub-task, specifically in the area of high-resolution coastal physical/ecological numerical modelling.

Ideally, biological models would have a mechanistic component based on first principles of species-specific planktonic life cycles and physiology. However, insufficient data for many species have hampered the development of robust, mechanistic models. As a result, many studies have focused on empirical or statistical methods to parameterise the unique relationships between environmental factors such as nutrients and temperature and the individual species responses within a defined geographic region (Anderson *et al.*, 2009; Kudela *et al.*, 2010). As ecosystem models move toward more accurate estimates of nutrients and chlorophyll biomass in the marine environment, their utility for forecasting species-specific dynamics, in particular HABs, becomes more realistic.

The capability of coupled physical-biological models can be further extended through use of empirical methods for defining environmental conditions strongly associated with target taxa. This approach has been implemented for the California Current System to hindcast and ultimately predict *Pseudo-nitzschia* blooms (Anderson *et al.*, 2009, 2011). A critical aspect of merging HAB dynamics with the Global Operational Ocean-Forecasting Network is that it is not sufficient for operational models to realistically predict physical and bulk biological (e.g., chlorophyll) parameters if the desire is to predict species-level ecosystem impacts; it is also not necessary to model species within operational modelling frameworks. Rather, it is critical for HAB researchers and managers to coordinate with the operational forecast effort to ensure that the model data are applied appropriately, and to provide feedback on how best to improve and utilise the coupled HAB/oceanographic models.

Information Technology Systems

An essential component of the observing system is the geo-spatial information system(s) that allows users to access the products they need, ideally in a very simple and intuitive manner. Such IT systems also act as the end-point integrators of many streams of complex data and information: it is the relatively simple, powerful, value-added products provided by these systems that represent the real value, not the underlying data in itself (European Commission Green Paper, 2012). Key aspects of these systems are the need for open-source, interoperable, and modular structures (Leadbetter *et al.*, 2013), and the need for simplicity of access. The ChloroGIN initiative (Sathyendranath *et al.*, 2010), with an existing distributed global dissemination network, is expected to play a major role. The HAB-related information system component is expected to make a contribution to the larger SB-01-C1: Sustained Ocean Observations sub-task.

Conclusion

HABs have huge economic and societal consequences around the globe, in both marine and freshwater systems. There is clear evidence that these impacts are increasing and becoming more costly. The availability of geo-spatial information on such blooms is of considerable value. The Blue Planet initiative offers a mechanism to greatly increase the availability of such information at relatively low cost, primarily through consolidation of existing international initiatives under a globally integrated strategic focus. Strong leadership from the Blue Planet initiative will facilitate the stated goal of building a global community to develop and maximise the use and societal benefits of an integrated HAB observation and prediction system.

CHAPTER NINE

LIVING MARINE RESOURCES: HARVESTING, ASSESSMENT AND MANAGEMENT

CARA WILSON AND JEFFREY POLOVINA

Introduction

The term "fisheries" encompasses not only commercial fish stocks, but also all living marine resources (LMR). There are three distinct aspects of fisheries: harvesting, assessment and management, which all have different goals. Harvesting efforts focus on increasing the catch-per-unit-effort (CPUE), that is optimising methods of finding, and more efficiently harvesting, fish. Traditionally, assessment involves both the species and its habitat. Stock assessments estimate either the total population or the total biomass of a fisheries stock in a given region, whereas habitat assessments characterise the environmental conditions favoured by a species. Fisheries management uses both stock assessments and habitat assessments to set harvesting limits and guidelines on commercial stocks to maintain sustainable exploitation and to develop regulations to help populations of protected and endangered species recover.

In the last half century, the world fish harvest has increased more than four-fold, reaching 90 million tons in 2000 (FAO Fisheries Department, 2004). At the same time, the numbers of overexploited and depleted stocks have risen, increasing the pressure on fisheries resources. Better management and understanding of fisheries are needed to maximise the utility of the current resources and to ensure their sustainability into the future. However, these issues are complicated by the significant inter-annual variations that occur naturally in fish population, and teasing these apart from fluctuations caused by anthropogenic effects is not trivial (Kendall and Duker, 1998; Bakun and Broad, 2003).

There are two primary ways that satellite data are used within fisheries. One is to find populations, usually a commercial fish stock to increase CPUE, but also in some cases for conservation. A second application is characterising and monitoring the habitat that influences LMRs. Satellite data provide an environmental context within which to examine these issues, by measuring parameters of the habitat and ecosystems that influence marine resources, such as upwelling, HABs, seasonal transitions and El Niño events.

Of the many types of data available from satellites, ocean colour data are particularly important, since ocean colour is the only remotely-sensed parameter that directly measures a biological component of the ecosystem. Satellite-derived chlorophyll data provides an index of phytoplankton biomass, which is the base of the oceanic food web. The relationship between satellite chlorophyll and a specific fish stock depends upon the number of linkages between phytoplankton and the higher trophic levels. There can also be spatial disconnects between satellite measurements of the ocean surface and demersal and deep-water species. Nonetheless, chlorophyll provides a key metric to measuring ecosystems on a global scale. Satellite chlorophyll measurements are the primary component in algorithms to calculate the primary productivity (PP) of the ocean. Global PP measurements, in conjunction with fish catch statistics and food web models, such as shown in Figure 9-1, can be used to estimate the carrying capacity of the world's fisheries. Studies suggests that coastal systems are at or beyond their carrying capacity (Pauly and Christensen, 1995), which is cause for concern as the bulk of the world's fish catch comes from coastal areas.

Links to Societal Benefits

Operational Fisheries (Harvesting)

Locating and catching fish is becoming more challenging as fish stocks dwindle and move further offshore, thus increasing the search time, cost and effort. Satellite data can help to increase CPUE by identifying oceanographic features that are often the sites of fish stock congregation and migration (Laurs *et al.*, 1984; Fiedler and Bernard, 1987; Chen *et al.*, 2005). As long as there are adequate catch limitations in place, and they are followed, increasing the CPUE is not incompatible with maintaining a sustainable fishery.

Both satellite ocean colour and SST data have been used to increase fishing efficiency. SST and ocean colour often have similar patterns as temperature, nutrients and chlorophyll are usually correlated. SST can also be an important factor for determining potential fishing grounds since different fish species have different optimal temperature ranges. Generally, SST data have been used more often than ocean colour data in fisheries applications, because SST data have been available for longer, have been more freely available in (near) real time, and several measurements can be obtained per day for a given location.

Satellite data must be available in near-real time in order to be used for increasing CPUE. In some countries, such as Japan and India, the national fisheries agencies are actively involved with helping increase the efficiency of their fishing fleets, whereas in other countries, such as the US, this is a commercial enterprise. The Indian National Centre for Ocean Information Services (INCOIS) uses data from the Indian ocean-colour satellite to make maps of potential fishing zones (PFZ), which are freely disseminated throughout coastal India by various venues. Studies on the effectiveness of the PFZ advisories have suggested that they have helped reduce search time by up to 70%, and have significantly increased the CPUE (Solanki *et al.*, 2003; Zainuddin *et al.*, 2004).

Stock Assessment

While satellite remotely-sensed data are now widely used in operational fisheries, their use in stock assessments is just beginning (Koeller *et al.*, 2009a). The incorporation of environmental data of any kind into stock assessment models has rarely been achieved successfully, for several reasons. First, assessments have traditionally taken classical single-species approaches which deal only with the numerical population dynamics of the stock under review. Second, the environmental factors forcing changes to the stock are complex, poorly understood and difficult to measure. Third, radical changes to stock assessment methodology, such as incorporating environmental data, would disrupt the inter-annual time series of a stock population.

However, the advent of the 'ecosystem approach to fisheries' has given new impetus to include environmental data as an integral part of the assessment process. It is particularly important to understand the factors determining recruitment of commercially important fish and shellfish stocks, for two reasons. First, the adverse effects of fishing cannot be separated from 'normal' environmentally-driven changes unless the latter are thoroughly understood. Second, environmental factors modify underlying stock-recruitment relationships. Until recently, defining stock-recruitment relationships and identifying the environmental factors modifying them have been the 'holy grail' of fisheries research, largely un-resolvable with traditional oceanographic methods because of the complex processes involved. Satellite data, particularly ocean colour, SST, and altimetry have now made these objectives achievable.

A long-standing hypothesis in fisheries has been that recruitment success is tied to the degree of timing between spawning and the seasonal phytoplankton bloom (Cushing 1969, 1990). This hypothesis has been difficult to prove or disprove with traditional shipboard measurements that have limited spatial and temporal resolution, but with satellite ocean colour data, inter-annual fluctuations in the timing and extent of the seasonal bloom can be clearly seen. The timing of the spring bloom determined from satellite ocean colour on the Nova Scotia Shelf was compared with available *in situ* data on larval survival of haddock, and successful year classes of haddock were associated with exceptionally early spring blooms of phytoplankton (Platt *et al.*, 2003). This study demonstrates that it can be possible to separate ecosystem-associated variability in fish stocks from other components, such as human exploitation or predation effects.

Habitat Assessment

Coral reef ecosystems support a high diversity of coral, fish, and benthic species. Coral reefs are sensitive to their environment (temperature, light, water quality and hydrodynamics), and as a result of both direct anthropogenic (e.g., mechanical damage) and climate impacts (Kleypas *et al.*, 2001), they are among the most threatened coastal ecosystems worldwide (Pandolfi *et al.*, 2003; Hoegh-Guldberg *et al.*, 2007). With the capability of providing near-real-time synoptic views of the global oceans and the ability to monitor remote reef areas, satellite remote sensing has become a key tool for coral reef managers and scientists (Mumby *et al.*, 2004; Maina *et al.*, 2008; Maynard *et al.*, 2008). Since 1997, NOAA has been producing near-real-time, web-accessible, satellite-derived SST products to globally monitor conditions that might trigger coral bleaching from thermal stress (Strong *et al.*, 2006). These products provide an effective early warning system globally, but are not always accurate in predicting the severity of a bleaching event at a regional scale (McClanahan *et al.*, 2007; Maynard *et al.*, 2008). CSIRO's ReefTemp project produces satellite-derived bleaching risk indices specifically for Australia's Great Barrier Reef (Maynard *et al.*, 2008).

Ecosystem Management

The availability of global coverage of SST and sea-surface chlorophyll from satellites has been instrumental in classifying marine ecological regions. The challenge, however, is in interpreting the relationship between the fundamental satellite measurement, i.e., the amount of chlorophyll in the surface ocean as measured by ocean colour data, and the higher trophic levels (Figure 9-1).

In the late 1980s, field programmes monitoring monk seal pup survival, sea bird reproductive rates, and reef fish densities in the North-western Hawaiian Islands indicated ecosystem changes had occurred. Data from SeaWiFS, the first ocean colour sensor launched in 1997, identified the Transition Zone Chlorophyll Front (TZCF), a separation between the cool, high surface chlorophyll, vertically mixed water on the north and the warm, low surface chlorophyll, vertically stratified subtropical water on the south (Polovina *et al.*, 2001). In some years the TZCF remains north of these northern atolls throughout the year, while in other years the TZCF shifts far enough south during the winter to encompass these atolls with higher chlorophyll water. The ecosystem of the northern atolls is more productive after the TZCF is more southerly located relative to its long-term winter position. Specifically, during a winter when the TZCF was shifted south of its average position, monk seal pup survival two years later increased (Baker *et al.*, 2007). The two-year time lag probably represents the time needed for enhanced primary productivity to propagate up the food web to monk seal pup prey. Should management action, such as a head start programme, be developed to improve pup survival, a two-year forecast based on satellite ocean colour can be used to predict the years when low survival is likely.

With fewer than 400 individuals left, the north Atlantic right whale (*Eubalaena glacialis*) is one of the most endangered whale populations (International Whaling Commission, 1998; Kraus *et al.*, 2005). This population spends much of its time in United States and Canadian waters, with the winter calving grounds off Florida, Georgia and South Carolina and feeding grounds in the Gulf of Maine. The recovery of this population is limited by high mortality, especially due to ship strikes and entanglements in fishing gear. The current management strategy involves limiting adverse impacts by requiring modifications to fishing gear or vessel speeds in regions and time periods when whales are likely to be present. Thus, all management options require knowing when and where whales are likely to be. A new approach to locating right whales combines synoptic information from satellites with a model of the right whales' main prey, the copepod *Calanus finmarchicus*. High numbers of whales are typically found in regions of high copepod concentrations (Pendleton *et al.*, 2009). Many important rates in the life cycle of *Calanus*, such as egg production and egg development times, can be estimated using satellite data. By combining the rate information derived from satellite data with reconstructions of the ocean currents from a computer model, estimated maps of *Calanus* abundance can be produced and related to right whale distributions (Pershing *et al.*, 2009a; Pershing *et al.*, 2009b). An initial test of this system forecasted that, due to the cold winter in 2008, the *Calanus* population would be delayed, and that whales would arrive on their main spring feeding ground east of Cape Cod three weeks later than normal. While a full analysis of the data is underway, it appears that the whales arrived close to when the model predicted. These forecasts are currently being expanded temporally and spatially, and will soon be able to incorporate observations of both copepods and whales.

A pelagic longline fishery based in Hawaii occasionally catches several species of sea turtles, with the threatened loggerhead sea turtle (*Caretta caretta*) historically accounting for the majority of the turtle by-catch. Since 1997, Argos-linked transmitters have been attached to loggerhead sea turtles caught and released by longline vessels (Polovina *et al.*, 2000), in order to characterise migration and forage areas of loggerheads, with the aim of ultimately spatially separating the fishery from the loggerheads. The use of satellite SST, ocean colour, altimetry and wind data have all been important in defining the oceanographic habitat of turtles within the North Pacific (Polovina *et al.*, 2000; Polovina *et al.*, 2004; Polovina *et al.*, 2006; Kobayashi *et al.*, 2008), allowing determination of seasonal habitat maps (Kobayashi *et al.* 2008). By combining this information with environmental satellite data, and fisheries and fisheries by-catch data, it is now possible to predict the locations of areas with a high probability of loggerhead and longline interactions (Howell *et al.*, 2008). In 2006, NOAA launched an experimental product called TurtleWatch, which uses satellite oceanographic data to map, in near real time, areas with a high probability of loggerhead and longline interactions, so that fishers can avoid them. The TurtleWatch tool is generated and distributed daily and is provided to fishers on board, via the GeoEye commercial fisheries information system.

Aquaculture

Ocean colour may be used to detect environmental conditions affecting shellfish growth, such as phytoplankton biomass, temperature and turbidity. Various high-resolution commercial satellites have been used for aquaculture applications, including mapping mussel farms (Alexandridis *et al.*, 2008), and site selection for near-shore aquaculture sites has been routinely based on the use of multispectral images from high spatial-resolution sensors (e.g., Landsat, Spot). Medium-resolution satellite data can be used for open ocean applications. For example, site selection of sea bream and sea bass cages near the Canary Islands (Spain) made extensive use of SST data for the identification of suitable culture temperatures in the region (Pérez *et al.*, 2003).

Conclusion

Satellite data characterise oceanic properties of habitat and ecosystems that influence LMR at spatial and temporal resolutions that are impossible to achieve any other way. The high spatial resolution provides an important geographical context for interpreting other data and results. The daily-to-weekly temporal resolution allows for effective detection, identification and monitoring of many oceanic features and permits the extraction of value-added products such as the timing of seasonal events. Time series of science-quality satellite data are needed to understand linkages between climate and ecosystems, and to characterise and monitor ecosystems as part of an ecosystem-based approach to fisheries management. For example, satellite data can be used to observe changes in the timing of the spring bloom that can affect recruitment (Platt *et al.*, 2003), to classify the productivity of the oceans (Sherman *et al.*, 2005), to detect inter-annual differences in the frontal structures that are important to fisheries (Bograd *et al.*, 2004; Polovina *et al.*, 2001) and to map the spatial extent of the ocean experiencing lower productivity during an El Niño event (Wilson and Adamec, 2001). Near-real-time satellite data are needed to optimise sampling for fisheries survey cruises for management and stock assessment and also to increase the efficiency of fishing efforts.

Acknowledgments

This paper is dedicated to **David G. Foley,** a dear colleague who passed away Dec 8, 2013. Since 1996 Dave worked for NOAA's CoastWatch program, making near real-time satellite data more easily accessible to a wide variety of end users. Dave led the field in incorporating satellite data into ecosystem and marine resource management. Dave was generous with his time, knowledge and code, and helped many scientists better understand the distribution and behaviour of a wide range of animals, including salmon, albatross, squid, sharks, whales, marine turtles and albatross. RIP Dave. http://www.forevermissed.com/david-g-foley

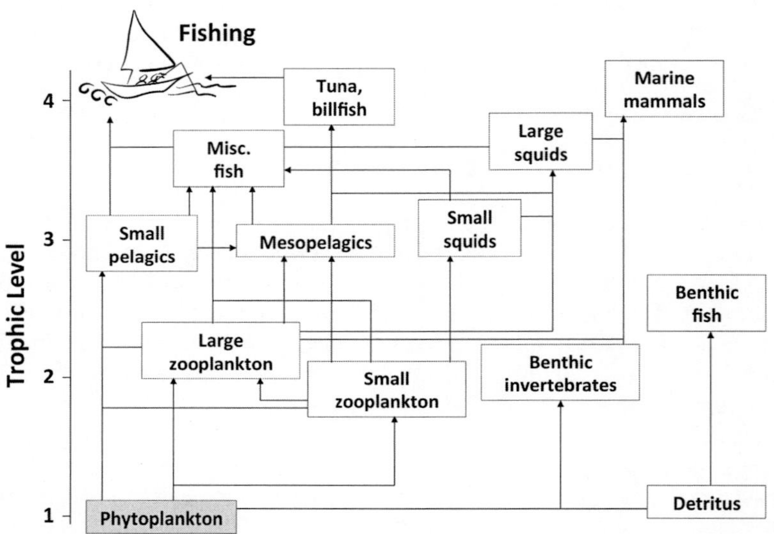

Figure 9-1: Simplified oceanic food web, showing the varying complexities in the linkages between phytoplankton, which is measured by satellite ocean colour data, and higher trophic levels. Modified from Pauly and Christensen (1993)

CHAPTER TEN

PHYTOPLANKTON PHENOLOGY AS AN ECOLOGICAL INDICATOR FOR THE PELAGIC SYSTEM IN THE OCEAN

LI ZHAI, TREVOR PLATT AND SHUBHA SATHYENDRANATH

Introduction

Phytoplankton phenology is defined as the seasonal phytoplankton cycle, in which the dominant event is the spring bloom. The inter-annual fluctuations in the phase of the cycle are important for the marine ecosystem. Phytoplankton phenology, as determined by ocean-colour radiometry, has proved to have a wide variety of applications, including fisheries management. For example, the timing of the spring bloom has been shown to be an extremely important indicator for the survival of Haddock fish larvae (Platt *et al.*, 2003, as summarised in Chapter 9). It influences the hatching times of northern shrimp (Koeller *et al.*, 2009b), and is related to the breeding success of marine seabirds (Borstad *et al.*, 2011). These findings support the match or mismatch hypothesis proposed by Cushing (1975). Cushing's hypothesis contends that there is an overlap in seasonal cycles of larvae production and biomass of phytoplankton, where larvae have a food supply adequate for survival. Where this is not so, larvae are vulnerable to death by starvation. Cushing's hypothesis is valuable in showing how a fish stock can respond to climate variation and to climatic trends (Cushing, 1975).

This chapter synthesises the concept of phytoplankton phenology by way of showing examples, including the characteristics of spring blooms derived from ocean-colour remote sensing and their response to environmental factors in the North Atlantic (Zhai *et al.*, 2011; Zhai *et al.*, 2012). A current limitation is the relatively short length of the time series that can be derived directly from remote sensing, compared with the data record for fisheries (Stuart *et al.*, 2011). Because phenology has a physical basis (onset of stratification), one possible remedy is to reconstruct through retrospective physical analyses, the onset of stratification and its relation to phytoplankton phenology, with a view to making a longer record of phenology for research on the connections between phytoplankton phenology and fisheries. We will demonstrate one example of retrospective estimation of phytoplankton inter-annual variability from the central North Atlantic Ocean (Zhai *et al.*, 2013). Societal benefits of ocean colour for fisheries management and climate change study will also be demonstrated in this chapter.

Phytoplankton phenology in the North Atlantic

The spring bloom is the dominant event in the seasonal cycle of phytoplankton and inter-annual fluctuations in phase are important for marine ecosystems (Platt *et al.*, 2007). Satellite ocean-colour data can be used to objectively construct the timing of the spring bloom initiation, the timing of the bloom peak and also the duration of the bloom. A shifted Gaussian function of time was used to fit the time series of SeaWiFS 8-d chlorophyll concentration. Figure 10-1b shows example time series of satellite chlorophyll at several stations. It illustrates two types of time series: short blooms with high peaks, and longer, less-intense blooms. The timing of the spring bloom initiation has a general latitudinal trend, with later blooms at higher latitudes. But this trend is often modified by regional physical forcing of vertical mixing.

Scotian Shelf and adjacent slope water

On the Scotian shelf (see Figure 10-1a), the timing of the spring bloom initiation varied spatially by about 2 months, with earlier blooms on the middle than on the eastern and western Shelf (see Figure 10-1b). The rapid shoaling of the mixed-layer depth from February to March triggered the early spring bloom on the middle Scotian Shelf. We computed vertically-averaged light in the mixed layer at the time of bloom initiation, an essential property in Sverdrup's theory, to reconcile the differences in bloom timing. This quantity has a narrow range of 15 ± 5 Wm^{-2}, compared with its annual cycle. The duration of the spring bloom has a strong across-shore gradient, with shorter duration on the shelf and longer duration in the slope water. The position of the strongest gradient coincides with the 95th percentile (northern limit) of the shelf-slope SST front position. The longer duration in the slope water is likely related to the continuous supply of nitrate to the mixed layer from beneath, because of the enhanced vertical and horizontal mixing in the frontal region. The amplitude of the spring bloom shows higher values in shelf water than in slope water and the Gulf Stream, and generally decreasing amplitudes from the eastern towards the western Scotian Shelf. The bloom amplitude and nitrate concentrations displayed similar spatial pattern on the Scotian Shelf, and the correlation coefficient between them is significant (r=0.87, Zhai *et al.*, 2011).

Iceland shelf and adjacent water

The waters around Iceland (see Figure 10-1a) are oceanographically complex, affected by the warm and saline Atlantic water, the cold and fresh Arctic, and colder Polar water. The boundaries between water masses fluctuate with time, making environmental conditions around Iceland variable (Zhai *et al.*, 2012). The annual chlorophyll cycle around Iceland (see Figure 10-1b) was generally characterised by a pronounced spring peak in all provinces, with higher bloom amplitude on the Iceland Shelf than in the deep water. In comparison, there are usually two well-developed spring and autumn blooms in temperate latitudes of the Northwest Atlantic. But local physical forcing often modified this trend. For example, a set of minor bloom peaks occurred in the Atlantic water south of Iceland and two blooms were well developed on the southern Iceland Shelf.

The timing of the spring bloom initiation varied by one month between different regions around Iceland (Figure 10-1b). The spring bloom started in April on the northern Iceland Shelf, whereas the bloom was delayed until mid-May in the Atlantic water and southern Iceland Shelf. The differences in the timing of bloom initiation are caused by the stability of the water column. The hydrographic structure and the mixed-layer depth are significantly different between the Arctic and the Atlantic waters; Arctic water is generally cold and fresh and has a stable surface layer in comparison with the Atlantic water. In winter, the Atlantic water south of Iceland has a deep mixed-layer depth of 700 m, whereas the Arctic water north of Iceland has a shallow mixed layer of 150 m. In spring, the melting sea ice and fresh water run-off promote the rapid shoaling of the mixed layer in the Arctic water, resulting in the earliest bloom there. To explain the differences in bloom timing, we computed the averaged light in the mixed layer at the time of the spring bloom initiation, and found that the mean value was 14 Wm^{-2}, which is similar to the value estimated in the temperate Northwest Atlantic.

Central North Atlantic

In the central North Atlantic, the seasonal cycle of phytoplankton is dominated by a single bloom in the spring. The spring bloom (see Figure 10-1b) began in mid-March, reached peak amplitude in May and lasted until August. The timing of the spring bloom initiation varied between mid-March and late April from year to year (Zhai *et al.*, 2013). Recent studies in this area (Zhai *et al.*, 2012; Taylor and Ferrari, 2011; Mahadevan *et al.*, 2012) suggest that an underlying property controlling the initiation of spring blooms is vertical stratification. It was found that the inter-annual variability of the bloom initiation in the central North Atlantic was negatively related to the North Atlantic Oscillation (NAO) index and positively correlated with the SST (Zhai *et al.*, 2013). There are two competing mechanisms that affect the mixed-layer depth in the region. In a negative (positive) NAO year, the weak (strong) vertical mixing induced by weak (strong) westerly winds causes a shallow (deep) mixed-layer depth, whereas the strong (weak) northward transport of warm and salty North Atlantic water promotes weak (strong) stratification and thus a deeper (shallower) mixed-layer depth. The nominal relationship found between NAO, SST and bloom initiation suggests that horizontal transport plays a more important role than vertical mixing in affecting the onset of the spring bloom.

The statistical relationship between the timing of the spring bloom initiation and NAO and SST can be used to estimate the bloom initiation in the past using readily available physical variables. Zhai *et al.* (2013) showed that adding previous data to the regression model improves the correlation between estimated and observed bloom initiation. A possible explanation is that the mixed-layer depth is affected by the water transport in the previous year. The extrapolated initiation of the spring bloom showed later blooms in the mid-1980s and earlier blooms in the 1990s, which was captured by a biogeochemical and ocean ecosystem model (Henson *et al.*, 2009).

Links to societal benefits

Marine ecosystems provide goods and services such as fisheries, provision of energy, recreation and tourism, nutrient dispersal and cycling and coastal protection (Fischlin *et al.*, 2007). Climate change, overpopulation, endangered species, water quality and national sovereignty are important problems facing society, with climate change the priority issue (IOCCG, 2008). Ocean colour remote sensing provides the only way to monitor the ocean ecosystem on synoptic scales. It can help us to cope with these societal issues, and thus be beneficial to society.

Fisheries Management

The societal benefits of phytoplankton phenology, as derived by satellite ocean colour, for fisheries management were foretold by Sir Alister Hardy, who, in his monograph on the world of plankton remarked, "the changes of fortune in fisheries, [...], must have their causes in the natural world; the relation between cause and effect, however, will only be understood when we have an unbroken record of the changes in the sea month by month over a very wide area for very many years. To solve our problems we must have a service of information from the sea like the meteorological observations from the air ... Such a service, moreover, might not only help us to trace cause and effect; it might allow us to infer the probable course of events to come. [...] be of value in guiding fishing boats to more economic fishing" (Hardy, 1959).

One of the most promising approaches to implement this line of thinking might be to analyse the effect of the strong seasonal signal in chlorophyll concentration retrieved from satellite ocean-colour data. For example, the inter-annual variability in the timing of the spring bloom peak has helped to explain fluctuations on the survival of larval haddock fish (*Melanogrammus aeglefinus*) in the Northwest Atlantic (Platt *et al.*, 2003). In the two years (1981 and 1999) which have exceptional haddock year-classes, the spring blooms occurred unusually early. An early bloom for the haddock with an extended spawning period might prevent fewer of the total larvae from starving to death.

The northern shrimp (*Pandalus borealis*) is currently one of the most important wild fishery resources in the world. Egg hatching of the northern shrimp coincides with the spring bloom at all latitudes within its wide geographical range from Gulf of Maine (42° N) to Svalbard (78° N) (Koeller *et al.*, 2009a). The survival of northern shrimp larvae in the Gulf of Maine is further compared to the offset in timing of the spring bloom peak and the median hatch date. The relationship suggests that shrimp survival is better when the hatch lags the bloom onset by about 40 days (Koeller *et al.*, 2009b). These results could have direct implications for fishery management of northern shrimp in the Gulf of Maine, because the fishery is carried out almost exclusively on egg-bearing females.

Climate change

Detecting long-term changes in phytoplankton biomass is an example of how ocean colour remote sensing can be used to improve our understanding of the impacts of climate change on marine ecosystems. Antoine *et al.*, (2005) combined 8 years of Coastal Zone Colour Scanner (CZCS) and 5 years of SeaWIFS ocean-colour data to identify long-term trends. Although the world ocean average chlorophyll concentration increases by about 22% over the past two decades, some regions of the ocean display declining concentrations. Using eight years of SeaWiFS data, Polovina *et al.* (2008) observed that the areas of low surface chlorophyll waters in oligotrophic provinces have expanded at average rates of 0.8 to 4.3% per year. The oligotrophic gyre in the North Atlantic expands most rapidly at 4.3% per year. The expansion of low chlorophyll water is consistent with global warming scenarios owing to increased vertical stratification in the mid-latitudes.

Phytoplankton groups derived from SeaWiFS imagery can also be used to assess the response of marine ecosystems to climate (Devred *et al.*, 2009; Alvain *et al.*, 2013). Devred *et al.* (2009) assembled a 10-year time series of ecological provinces for the Northwest Atlantic Ocean using ocean colour data. Significant trends in diatoms and phytoplankton biomass vary according to the province concerned, emphasising how the physical and biological changes are correlated at large spatial scales. Large-scale shifts in the frequency of diatom dominance between positive and negative phases of North Atlantic Oscillation index are detected by ocean colour in the high latitude regions of the North Atlantic, and these shifts are also observed by the Continuous Plankton Recorder data (Alvain *et al.*, 2013; Zhai *et al.*, 2013).

Phytoplankton takes up CO_2 through photosynthesis, and a portion of it sinks from the surface layer as dead organisms and particles to a deeper layer which has no contact with the atmosphere (Denman *et al.*, 2007). Satellite ocean colour allows us to estimate phytoplankton production (the conversion of inorganic carbon to organic carbon through photosynthesis), on a global scale. The annual uptake of CO_2 in all oceans estimated from ocean-colour data is ~ 50 gigatonnes carbon per year, which is close to the total primary production on land (Longhurst *et al.* 1995), and provides essential information for initialising and validating ocean ecosystem models used to study global carbon cycles and climate change (Cox *et al.*, 2000).

Conclusion

This chapter used three examples from the North Atlantic to demonstrate that phytoplankton phenology is strongly linked to the environmental forcing of nutrient supply, averaged light in the mixed-layer depth, SST and NAO. Because phytoplankton exerts bottom-up control on the upper trophic levels of marine food webs, the retrospective phenology derived from the physical forcing could be used for the development of knowledge-based marine ecosystem management, and could be used to develop an ecological partition of the waters off Eastern Canada for use in fisheries applications.

Chapter Eleven

Importance of Time-Series Studies: The Latin-American Antares Network

Vivian Lutz

Introduction

Continuous studies of oceanographic properties allow scientists to understand the response of ecosystems to natural cycles and to anthropogenic effects. Ocean-integrated observations from long-term time series in different sites are needed to obtain a global view of changes affecting the world's ecosystems. The Antares network was created in 2003 with the aim of joining coastal time series of bio-optical and oceanographic studies in countries around Latin-America. In 2006, this idea was expanded to other regions of the world, creating the 'Chlorophyll Globally Integrated Network (ChloroGIN)' (see Chapter 12). Antares develops three main tasks: 1) *in situ* ship observations at local time-series stations; 2) remote sensing observations to expand the spatial and temporal coverage of the *in situ* information; and 3) capacity building to train young scientists from Latin-America in bio-optics and biological oceanography, as well as to facilitate the exchange of expertise among researchers in the network.

Antares was built on on-going projects in each of the participating countries, each with their own particular objectives, needs and availability of resources. Maintaining long-running oceanographic time series in developing countries represents a tremendous challenge since the budgets of the institutes and funding agencies is lower, and the cost of obtaining instruments and supplies is much higher, than in developed countries. The opportunities to obtain technical assistance and scholarships for students' training are also significantly lower. Despite the challenges, the network has achieved some important milestones.

Links to Societal Benefits

Earth's systems are being affected by human activities, including those leading to an increase in atmospheric (and oceanic) CO_2. Specific environmental changes which can be associated with anthropogenic effects include increases in global temperature, melting of ice caps, rising of sea level, decreases in biodiversity, and changes in species distribution. In turn, these ecosystem changes affect human society (health, food availability, livelihood, etc.). These effects may worsen as human population increases (at an estimated rate of about 1 million every 10 days). It is crucial to understand the type and magnitude of these changes in the Earth's systems so that appropriate decisions can be taken to avoid, or at least reduce, further deleterious effects and to design mitigation/adaptation strategies. Earth observations are critical to understanding how ecosystems work and how they react to external forcing.

The ocean plays a fundamental role in climate regulation, through both physical interactions and biogeochemical processes. Phytoplankton photosynthesis, for example, takes up atmospheric CO_2 and fixes it within living organisms, many of which ultimately die and sink to the bottom of the sea, where they may accumulate, thus locking carbon out of the system for a prolonged period of time.

Models to predict the effect of climate change must more fully describe and predict these marine processes, but modelling efforts to date point out the scarcity and gaps in data. Ocean observations, including environmental variables, biological components and biogeochemical fluxes are difficult and costly to make, however. Presently, observations can be made from a number of different platforms, including:

1. Satellite remote sensing, which offers a synoptic (ideally, global), high frequency (ideally, daily) and economic (strong initial investment provides large volume of data for years) view of several properties (e.g., temperature, chlorophyll concentration, wind) at the ocean's surface.

2. *In situ* buoys, which, although limited to particular sites in the ocean, provide higher frequency measurements (scales of minutes or less), can carry a larger variety of sensors than satellites (e.g., salinity, oxygen, fluorescence), and can measure properties at different depths within the water column.

3. *In situ* ship cruises, which cover limited spatial and temporal scales, are costly and laborious, but provide measurements of an even wider variety of properties, of the highest quality (used to

calibrate and validate automatic estimates by satellites and buoys) and at all depths. In addition, cruises allow the possibility to collect samples for biodiversity assessments (from bacteria to fish), and are still the only way of carrying out process studies (e.g., photosynthesis, grazing), which are essential inputs for modelling. Time-series measurements made at the Antares stations represent a valuable, sustained and long-term source of data, which can, in combination with satellite data, be used for climate change models.

In situ measurements

New time-series studies were incorporated in the network; there are presently sites in eight Latin-American countries: Argentina, Brazil, Chile, Colombia, Ecuador, Mexico, Peru and Venezuela (see Figure 11-1). In addition an important contribution in form of advisory support is provided by members from both the United States and Canada. An inter-calibration exercise on measurements of *in situ* chlorophyll-*a* concentration was carried out in 2005 (Lutz *et al.*, 2007). The outcomes of this first inter-calibration were to point out existing variations, and draw recommendations to improve the measurement of this key biological variable. Another important step towards improving the comparability of different types of measurements within the network was the participation of representatives of Antares sites in the 'Moving toward global intercomparability in a changing ocean: An international time-series methods workshop' carried out in November 2012 in Bermuda (Lorenzoni and Benway, 2013). A forum to maintain an active interaction among the 25-plus, ship-operated time series that participated in the Bermuda workshop was proposed, and will serve to advance the implementation of best-practice techniques. One of the highlights of this workshop was recognising the need for measurements from all over the world, and therefore the challenges of carrying out time series in developing countries were taken into account; e.g., recommendations for techniques to measure a certain property included not only the very best (more sophisticated and expensive), but also the acceptable best way to do this type of determination.

Remote Sensing

Antares implemented a system for processing and distributing satellite images and data, through a small-grant from the Inter-American-Institute for Global Change Research (IAI), SPG-II-026. The system adopted was one developed at the Institute of Marine Remote Sensing (USF, United States). This remote sensing information proved to be useful not only for network members, but for outside researchers as well, including fisheries biologists from different countries in Latin-America. This system is being updated and expanded at the moment.

Preliminary results combining both *in situ* and satellite data of SST and chlorophyll concentration at the different sites, showed that each one represents a different biogeographical province (Longhurst, 1998), with different trends in the anomalies of these properties during the period 1998-2007 (Santamaría-del-Angel *et al.*, 2010). This first integrated study is currently being completed.

International Cooperation

A proposal to develop an 'Assessment of marine ecosystem services at the Latin-American Antares time-series network' has been approved by the IAI, and will be undertaken over the next four years. This will be an unprecedented effort to integrate natural and social science studies, addressing the benefits to society provided by the ocean, phytoplankton in particular, in the Latin-American countries involved.

To foster capacity building, two regional courses were held in Brazil, and several students from Antares stations also attended training programmes at the 'NF-POGO Centre of Excellence in Oceanography in Bermuda'. All of these training programmes were sponsored by the Nippon Foundation and the Partnership for the Observation of the Global Oceans (NF-POGO). To promote the interaction among alumni, these organisations created the 'NF-POGO Alumni Network for Oceans (NANO)', and financed small regional projects for the alumni. A synergy between the Antares and the Latin-American NANO networks was sought in the formulation of the corresponding regional project, which has been running since 2012. The main aim of this project was to enhance the use of phytoplankton pigment analysis (by HPLC technique) in the *in situ* time series. The first phase included collection of samples at different Antares stations, and thanks to a generous agreement, these have been analysed at the NASA laboratory. The second phase includes training in interpreting the results obtained, as well as surveying the requirements to implement new pigment determinations at the different sites or improve existing ones (more information at: http://www.nf-pogo-alumni.org/Latin+ American+ Regional+Project).

Conclusion

Latin-America is a heterogeneous region not only in the natural realm, but also in the challenges that its societies and scientific research communities are facing. Increasing the interaction within the network and, developing common projects (such as the IAI Collaborative Research Networks '*CRN3*' programme), will help increase the types and quality of ocean observations in the region. The project to study ecosystem services at the Antares time series, will seek to integrate *in situ* information gathered at each of the nine field stations on changes in phytoplankton biodiversity, and the main environmental properties, which will be expanded to larger spatial and temporal scales through the combined use of remote sensing information and a regional modelling exercise. Furthermore, communication with representatives of the society (stakeholders and decision makers) will be carried out from the beginning to identify the main questions and possible factors of vulnerability in relationship to ecosystem services provided by phytoplankton. A methodology to estimate natural/socioeconomic indicators will be developed. A strategy of effective communication of the information provided by the indicators will also be investigated. This project will provide an unprecedented opportunity to link the natural information (on changes in the marine ecosystem) to the impact on ecosystem services, and hence on the well-being of society in the Latin-American region. The interaction of researchers from the natural and social sciences (e.g., oceanographers, economists, experts in ecosystem services, anthropologists) in such a project is expected to move ocean research an important step forward towards helping human society.

Continuous and high quality observations, on how the marine ecosystem is changing, are at the base of any plan to help in the mitigation/adaptation to climate change. A larger scientific effort is needed, from both developing and developed countries together, if we want to create a clearer picture of how our ocean is changing at the global level. National plans in the different countries, as well as international programmes such as NF-POGO, IOCCG, IOCCP, NASA, IMBER, and GEO through the new Blue Planet Task, are key players in facilitating this goal.

Chapter Twelve

The Chlorophyll Globally Integrated Network (ChloroGIN)

Steve Groom

Introduction

To create its vision of an international network of observations to assess the state of marine, coastal and inland-water ecosystems for the benefit of society, ChloroGIN promotes the use of in-water observations in synergy with ocean-colour and related satellite observations. ChloroGIN was initiated following recommendations of the "Plymouth Chlorophyll Meeting and Workshops (Extended Antares Network)" sponsored by the GOOS, GEO, IOCCG, Plymouth Marine Laboratory (PML) and POGO. The meeting took place on 18–22 September 2006 and was inspired by the Antares network (described in Chapter 11).

Objectives

The objectives of ChloroGIN, as noted on the ChloroGIN web site, are:

- "To provide time series of ecosystem indicators for the management of fisheries, aquaculture, coastal zones, and inland water bodies.
- To enhance *in situ* observations to characterise local conditions, including sub-surface water properties, for assessing water quality, ecological status and potential for human use. These *in situ* data are also essential for ground-truthing of satellite observations.
- To develop and disseminate value-added products needed for management by local authorities, regional and national governments.

- To develop tools and techniques for analysis of ChloroGIN products for assessing regime shifts, climate change and climate variability and the status of fisheries, ocean, coastal and inland water ecosystems, as a basis for policy implementation.
- To communicate ChloroGIN products of ecosystem state widely to potential users in society and in policy-making bodies.
- To help train and develop a new generation of scientists with the skills to develop and maintain the ChloroGIN vision.
- To achieve these objectives by means of an international network of both developed and developing countries.
- To liaise and cooperate with other Earth Observation groups nationally, regionally and internationally."

GEO is constructing GEOSS that is comprised of a number of building blocks, such as Architecture and Capacity Building and addresses a number of societal benefit areas, such as Ecosystems and Biodiversity (see Chapter 2). ChloroGIN addresses a number of these building blocks and societal benefits as described in the sections below.

The ChloroGIN portal (http://www.chlorogin.org) provides a description of the overall objectives, links to *in situ* data sources, and a number of portals to satellite Earth Observation (EO) data. Although the project does not have direct funding, its development is supported through a number of international projects inspired by the project. This chapter describes some recent and future developments of ChloroGIN.

Current status and links with GEO/Blue Planet

The ChloroGIN web site is maintained and developed by the Plymouth Marine Laboratory on behalf of the ChloroGIN project in support of GEO SB-01 Oceans and Society: Blue Planet, C2 Sustained Ecosystems and Food Security. The portal provides links to EO data, both archived and near-real time, from NASA, ESA and NOAA instruments, both globally, from data suppliers such as NASA and ESA, and locally with regional portals such as in South America and Africa (see Figure 12-1a).

The Africa portal has undergone significant development recently as a result of a European Commission funded project called EAMNet (Europe Africa Marine EO Network) that ran from March 2010 to June 2013. EAMNet established data production of ocean colour and sea-surface temperature data covering the entire coast of Africa at 1-km resolution (see Figure 12-1b). Clicking on any of the squares gives the option of retrieving data from either the University of Cape Town or Plymouth

Marine Laboratory; for example, clicking on the square covering the Mozambique channel highlights that data are available regionally from PML and the University of Cape Town (UCT) and from the European Commission's Joint Research Centre (JRC), who operate a pan-African data service. Clicking on the PML data source takes the viewer through to a PML portal where data can be viewed and downloaded (see Figure 12-1c). The Africa portal was extended in 2011 to provide coverage over the western Indian Ocean in support of the marine and coastal theme of the African Monitoring of Environment for Sustainable Development (AMESD) project and data production will continue during the follow-up Monitoring of Environment and Security in Africa (MESA) programme.

A key aim is providing data to end-users even in areas with poor internet coverage. This was assisted in EAMNet by the broadcast of data via the EUMETCast/GEONETCast system that uses Digital Video Broadcast over satellites so that users with their own receivers can obtain an on-going stream of data. EAMNet installed five GEONETCast systems to complement the five installed in an earlier EC project, DevCoCast; however, the data can be received by all the receiving stations in Africa. Interestingly, the installation of a receiver at the Makerere University, Kampala, Uganda, encouraged *in situ* sampling in the nearby Lake Victoria as ground-truthing for the satellite data.

EAMNet supported another ChloroGIN objective, namely to train and develop the next generation of scientists, and capacity building activities were undertaken at different levels including:

- The development of an Africa-focused module on marine EO that was presented either as part of M.Sc. programmes or as stand-alone two week courses at: the University of Cape Town, South Africa; University of Ghana, Ghana; University of Dar-es-Salaam, Tanzania, and at El Jadida, Morocco, as part of the JRC Satellite oceanography course, which EAMNet co-sponsored. In all, over 100 students from 16 different African countries were trained in a consistent manner.

- A fellowship programme, which enabled researchers and post-graduate students to visit centres of excellence in marine science, either in Africa or in Europe. Open fellowships were available for any scientist from any country to apply whereas targeted fellowships were available only to EAMNet consortium members. In all, 27 fellowships were undertaken by scientists from 11 African countries and involved topics such as: EO-based ecological indicators, fisheries resource management, oil spill monitoring, water quality, *in situ* data analysis, and numerical models.

Beyond the marine domain, in 2012, a ChloroGIN-Lakes portal was established that provided access to full resolution (300m) products from the ESA Medium Resolution Imaging Spectrometer (MERIS) instrument on the ENVISAT spacecraft (Figure 12-1d,e). This was done as an "end to end demonstrator" in support of the GEO WA-01 Integrated Water Information (including Floods and Droughts), C4 Global Water Quality Products and Services that explicitly mentions expansion of ChloroGIN to lakes. The demonstrator included example lakes in all continents with a focus on lakes where *in-situ* observations are available. Unfortunately, contact was lost with ENVISAT on 8 April 2012, since when there have been no images. There is no direct replacement for MERIS, but NASA's MODIS instrument has a very limited ocean colour capability at 500m resolution; an aim is to re-establish a demonstrator showing the capability of providing near-real time lake data in preparation for ESA Sentinel 3 missions that will carry the OLCI and provide global 300 metres resolution ocean colour data.

Future developments

A number of developments are on-going for ChloroGIN:

First, a link is being made to global products from the ESA Ocean Colour Climate Change Initiative project (OC-CCI) that is providing merged ocean colour data, at ~4-km resolution, derived from the SeaWiFS, MODIS-Aqua, and MERIS spacecraft.

Second, links are planned between ChloroGIN and a web-based data visualisation and analysis portal being developed by the EC-supported Operational Ecology (OpEc: http://marine-opec.eu) project that include *in situ* and EO data alongside numerical model outputs. The OpEc portal has dynamic tools that allow for fast visual comparison, as well as more complex summarisation and comparison operations (e.g., Figure 12-1f). Data will be displayed in map form (model fields, remote sensed data, error quantification fields, area-based summaries) and plots (*in situ* data, extracted data at points).

Third, a link is being constructed between the portal and the GEOSS Data-CORE a distributed pool of documented datasets with full, open and unrestricted access at no more than the cost of reproduction and distribution). The aim is to be able to access the wide range of resources, such as *in situ* data, that have been registered with GEO.

Finally, ChloroGIN will continue to expand coverage by linking with regional data providers' portals, in-line with the concept of regional centres of excellence providing input to a global activity.

Conclusion

ChloroGIN provides a framework for the development of activities associated with observation of marine ecosystems. It is contributing to GEO through Blue Planet and through ChloroGIN-Lakes, to the WATER societal benefit area. Future developments aim to expand coverage, increase linkages between the *in situ*, modelling and EO activities, and link further with GEOSS.

PART IV

OCEAN FORECASTING

CHAPTER THIRTEEN

THE GLOBAL OPERATIONAL
OCEAN-FORECASTING NETWORK:
GODAE OCEANVIEW

KIRSTEN WILMER-BECKER, MIKE BELL,
ERIC DOMBROWSKY AND ANDREAS SCHILLER

Introduction

Production of near real-time analyses and forecasts of the three-dimensional structure of the deep ocean's temperature, salinity, and currents started during the 1980s. Some groups (e.g., Derber and Rosati, 1989) were developing global systems with the aim of making seasonal predictions, whilst others (Thompson *et al.*, 1986) were focused on limited area, high-resolution systems aiming to predict the location of the "mesoscale" meanders and eddies within western boundary currents for Naval applications.

By the mid-1990s, as anticipated in part by Hurlburt (1984), several technological capabilities had matured sufficiently for high resolution global ocean prediction to have become feasible:

- Satellite altimeters, measuring the height of the sea surface (relative to the geoid), had been flown and achieved an accuracy of measurement sufficient to map the sea surface of the mesoscale eddies in near real-time across the global ocean (Le Traon *et al.*, 2009);
- Profiling floats had been developed during the World Ocean Circulation Experiment (WOCE) to the point where a global "Argo" system could be proposed to consist of 3000 floats sampling the ocean's temperature and salinity structure down to 2000 metres depth once every 10 days (Argo was initiated as a joint project of GODAE and CLIVAR);

- A number of community ocean models developed during WOCE, primarily for climate simulations, were available for ocean prediction (Chassignet *et al.*, 2009), and some first generation ocean data assimilation systems had been developed; and
- Computer technology had advanced to the point where eddy-resolving global simulations were starting to appear feasible.

The Global Ocean Data Assimilation Experiment (GODAE) was proposed by Smith and Lefebvre (1997) to support the groups leading the development of national ocean prediction systems. GODAE was conceived to be a finite life-time (10-year) demonstration of both the feasibility and the utility of high-resolution global-scale ocean predictions. It was led by an International GODAE Science Team (IGST) incorporating the key players in the teams developing the ocean prediction systems at the national level. The main outcomes from GODAE were captured in a Special Issue of the Oceanography magazine (Bell *et al.*, 2009). By this time, most of the members of the IGST had established global or regional operational ocean forecast systems producing analyses and forecasts of the meanders and eddies of the major world current systems, similar to those illustrated for the south coast of Africa in Figure 13-1.

At the end of GODAE, the members of the IGST were keen to sustain the networks and international coordination of activities established within GODAE, and GODAE OceanView (GOV) was proposed as a follow-on. This chapter gives an overview of the prediction systems developed by the participants in GODAE OceanView together with their associated societal benefits, services, and dependencies. It also outlines the international co-ordination activities planned by GODAE OceanView in 2009 and the progress made by its Task Teams over the last five years.

Ocean Forecasting at the National level

The operational ocean-forecasting systems developed at a national level over the last two decades are at the core of GOV. These systems routinely provide high-resolution real-time and hindcast, ocean analyses and forecasts for the global ocean and regional seas. Key elements of these systems include mechanisms to receive, exchange and quality control measurements from satellites and *in situ* platforms, forecast models and advanced data assimilation schemes, high performance computing facilities, and service delivery infrastructure. Table 13-1 indicates the domains and resolutions of the prediction systems operated by the groups participating in GOV. The model resolutions vary depending on the system and the area covered (global or regional). Horizontal resolutions range from $1°$ to $^1/_{10}°$ or higher reflecting the geospatial grid-spacing, while vertical resolution is set up either through *levels (between 40-75)* of z (height), σ (terrain following), isopycnic (density), or as hybrid (isopycnal-sigma-pressure combined) coordinate *layers (between 25-35).*

Most of the systems produce forecasts on a daily basis. Many of them are based at National Met Services, allowing them to share computing facilities and maintain close relationships with groups developing Numerical Weather Predictions (NWP), which greatly facilitates the development of coupled atmosphere-ocean prediction systems. Detailed information is available from the GOV website (https://www.godae-oceanview.org/documents/q/category/govst/system-reports/), with system reports being updated on an annual basis.

Links to societal benefits

A number of societal benefits can be achieved using the information generated by these systems. Predictions of near-surface currents are used to assist ship routing; responses to releases of toxic materials, oil, or chemicals through estimates of their drift and dispersion; and the determination of search boxes in search and rescue operations. Hindcasts and predictions of current profiles are used to improve the safety and efficiency of the oil and gas industry, both in its planning and operational phases. Short-range weather forecasts rely on accurate determination of ocean surface temperatures and heat content, and seasonal predictions require information about anomalies in the depth of the ocean thermocline. Information from these systems is also needed to interpret and place into context the events occurring in the broader marine ecosystem, such as the concentrations of marine life associated with frontal upwelling zones, the occurrence of algal blooms and eutrophication, and the life cycles of fish. The systems are also used to provide boundary conditions for shelf-sea and coastal monitoring and prediction systems.

Achievement of these societal benefits depends on a complex supply chain:

First, the predictions usually need to be tailored into products appropriate for specific segments of the market by teams (in either the public or commercial sector) with niche skills and expertise. These teams typically rely on the timely availability of generic predictions of known accuracy and high quality.

Second, the quality of the analyses and forecasts produced by these systems is heavily dependent on a continuous, near real-time, and sustainable supply of satellite and *in situ* observations of, for example, sea level, SST, ocean colour, sea ice, and surface waves and winds (Schiller and Brassington, 2011).

Third, in order to make good use of new technologies and research and development, and to better meet user needs, continuous improvements are made to the forecasting systems. GOV facilitates effective investment in research and development at the system level through international knowledge exchange and collaboration between system representatives. The developments made to the operational forecasting systems since 2009 include:

- Increased model resolution, inclusion of tides, improved surface forcing and surface fields, wave-current interactions, and progress with coupled model runs (physical/biological and ocean/atmosphere/ice) at many of the GOV systems;
- Assimilation of new observations including sea ice, salinity, tides, and biological observations;
- Realisation of observation impact studies, reanalyses, improved error covariance estimates and bias correction schemes, higher resolution regional predictions, and ensemble forecasting;
- Broadening of the systems to include biogeochemistry and ice predictions. Most systems are now working towards coupled physical-biological state estimation (both global and regional) and coupled ocean-wave-ice-atmosphere predictions.

International Cooperation

Like GODAE, GODAE OceanView is led by a Science Team (the GOVST) consisting of key players within the national operational ocean prediction centres. It has two main purposes. The first is to accelerate the improvement and exploitation of the systems at these centres through exchange of information and expertise. The second is to coordinate joint work to improve the:

- Assessments of forecasting system performance;
- Evaluations of the dependence of the forecasting systems and their societal benefit on the components of the observation system;
- Exploitation of the forecasting systems for greater societal benefit.

The plans and goals for the first 5 years of GOV are described in the GOV work plan (https://www.godae-oceanview.org/files/download.php?m =documents&f=120427154755-GOVWorkPlanUPDATEApr2012.pdf). During the planning phase of GOV, it was decided to establish two Task teams to work on the first two bullet points listed above. Recognising that the societal benefits from the ocean prediction systems would require joint work with other teams of experts, three additional Task teams were established to work jointly with the relevant communities developing the prediction of shelf seas and coastal waters, the analysis and prediction of open ocean marine ecosystems, and coupled, short-range, weather predictions. The following paragraphs describe the motivation behind, and progress being made, in the Task teams.

The **Intercomparison and Validation Task Team (IV-TT)** coordinates and promotes the development of the scientific validation and inter-comparison of national operational oceanography systems. This includes defining metrics and establishing specific global and regional inter-comparison experiments to gather information on how well the operational systems are performing. The real-time verification inter-comparison now has five participants providing model data for coordinated comparison with measurements for some or all of the following parameters: sub-surface temperature and salinity data for comparison with Argo floats, surface temperature data for comparison with *in situ* drifters, and sea-level anomaly data for comparison with satellite altimeters. A multi-model activity to assess the benefits prediction ensembles of surface fields by a number of operational centres is also in progress.

The IV-TT liaises with the Expert Team for Ocean Forecast Systems (ETOOFS) organised by JCOMM (Joint Technical Commission for Oceanography and Marine Meteorology) on operational implementation. It also collaborates closely with CLIVAR/GSOP (Climate Variability and Predictability/Global Synthesis and Observations Panel) on the inter-comparison of re-analyses. An on-going inter-comparison focused on the 2005-2010 period is targeting essential ocean variables such as ocean heat and salt content, sea level, and the Meridional Overturning Circulation. The outcomes will provide a methodology for longer period assessments and approaches for real time monitoring of the global ocean variability.

The **Observing System Evaluation Task Team (OSEval-TT)** focuses on evaluating and improving the understanding of the impact of observations on ocean model forecasts. It was formed by GODAE OceanView and the Ocean Observation Panel for Climate (OOPC) to provide evidence-based responses to agencies and organisations in charge of sustaining the global and regional ocean observing systems used for ocean monitoring and forecasting at short-range, seasonal, and decadal time-scales. This activity requires consistent protocols for observation impact assessment, tools for routine production of appropriate diagnostics, common sets of metrics for inter-comparison of results, and objective methodologies for observing system design and assessment activities.

Recently, this group performed a prototype Observing System Experiment (OSE), demonstrating the impact of different observing systems (Argo, XBT, TOA, Jason-2 and other altimeters, AVHRR) on a prediction system (Lea, 2012). The plan is to run the same experiment at other national systems to provide a more robust evaluation of the impact of the observations.

The **Coastal Ocean and Shelf Seas Task Team's (COSS-TT)** main goal is to work towards the provision of a sound scientific basis for sustainable multidisciplinary downscaling and forecasting activities in the world's coastal oceans and shelf seas. The team plans to help achieve a truly seamless framework from the global to the coastal/littoral scale. The main disciplines considered by the team are physics and the interaction between physical and biogeochemical processes. The TT is particularly interested in providing input to the choice of observations in coastal regions and establishing Observing System Simulation Experiments (OSSEs) along with OSEs to support this. Information about the systems developed by many national operational and research groups involved in coastal and regional modelling, as well as an overview of each group's objectives, products, and applications has been collected on the GOV website (https://www.godae-oceanview.org/files/download.php?m=docum ents&f=130205143030-COSSTTSITJan2013.xlsx).

The **Marine Ecosystem Monitoring and Prediction Task Team (MEP-TT)** is working to bridge the gap between present physical modelling and forecast capabilities and new applications in areas such as fisheries management, marine pollution, and carbon cycle monitoring. The team's goal is to define, promote, and coordinate actions between developers of operational systems and ecosystem modelling experts in order to integrate new models and assimilation components for ocean biogeochemistry and marine ecosystem monitoring and prediction.

The most recent activities focus on the evaluation of the impacts of the assimilation of physical data, and of satellite ocean colour and surface partial pressures of CO_2, on biogeochemical models. The Task team also plans to link closer to downstream applications (e.g., BGC monitoring and fisheries), and to work more closely with a number of international programmes (International Council for the Exploration of the Sea ICES, GOOS, Argo).

The **Short- to Medium-Range Coupled Prediction Task Team's (SMRCP-TT)** objective is to draw together international scientific and technical expertise in ocean, sea-ice, and wave prediction, and to seek collaboration with equivalent expert groups in atmospheric-land surface-hydrology prediction in order to accelerate the scientific and technical development of fully coupled prediction systems on short- to medium-range timescales.

The recently organised joint workshop with the Working Group for Numerical Experimentation (WGNE) of the WCRP (World Climate Research Project)) facilitated valuable international discussions and formed the basis for a review of the status, gaps and scientific challenges of developing coupled predictions. Publication of a white paper covering four topics relevant to coupled prediction (observations, physical parameterisation, dynamical modelling and data assimilation) is planned for 2014.

The Task teams and the GOVST are supported by the GODAE OceanView Project Office which maintains the GOV website, organises annual science team meetings and Task team workshops, and coordinates the organisation of major events such as symposia and summer schools. The Project Office fosters close links with the members of GOV and many international research groups. Recognising the development of national systems in China, India, and Brazil, the GOVST has broadened its membership to these countries in recent years. The GOVST also liaises closely with JCOMM's ETOOFS and has supported a student summer school (Schiller and Brassington, 2011).The Project Office and initiatives agreed to by the GOVST are supported by a Patrons Group, which consists of senior representatives of the national ocean prediction centres and representatives from other institutions (e.g., Space agencies) who are willing to support the work.

GODAE started 15 years ago as an experiment. Recognising that GODAE OceanView is now evolving towards a long-term international programme for Ocean Analysis and Forecasting, a formal review of GOV has recently been undertaken. This provided an independent assessment of the value of the GOVST and advice on how to improve its effectiveness. The findings from the review were very positive and its recommendations will be used to sharpen the focus of future coordination and to help to find a suitable long-term home for the programme.

Conclusion

Significant progress has been made in many countries in the implementation of ocean-forecasting systems and services. The GOVST has helped to accelerate this development through coordination of joint activities and sharing of expertise. There are many opportunities and challenges still facing GOV and each of its GOV Task Teams. Improved realisation of societal benefits from the existing capabilities is a particularly important challenge, one in which GEO may provide valuable assistance to GOV.

Table 13-1: Overview of GOV system characteristics, including country, model domain, and horizontal/vertical resolutions; horizontal resolution is categorised as follows: Low: $1° - {}^1/_2°$, Medium: ${}^1/_4° - {}^1/_{10}°$, High: ${}^1/_{10}°$ or higher. More detailed system characteristics can be found on the GODAE OceanView website: _https://www.godae-oceanview.org/science/ocean-forecasting-systems/_

System name Institute/ country	Global domain (horizontal resolutions)	Regional domains (horizontal resolution)
BLUElink CSIRO; Australia	Global (low)	Australian regions and Tasman Sea (high)
CONCEPTS EC, DFO, DND; Canada	Global (medium)	Arctic, North Atlantic, and Great Lakes (high)
ECCO JPL/NASA; USA	Global (low)	Tropical Pacific, Atlantic, Gulf of Mexico, and other regions (high)
ECMWF UK	Global (low)	N/A
FOAM Met Office; UK	Global (medium)	North Atlantic, Indian Ocean, Mediterranean (high)
HYCOM/NCODA (GOFS) NOAA, US-Navy, Universities; USA	Global (high)	N/A
INDOFOS INCOIS; India	Global (low)	Indian Ocean (high)

System name Institute/ country	Global domain (horizontal resolutions)	Regional domains (horizontal resolution)
Mercator Ocean France	Global (medium to high)	North Atlantic, Mediterranean Sea, Iberian Biscay, and Irish Seas (high)
MFS Italy	N/A	Mediterranean Sea (high)
MOVE.COM/MRI JMA/MRI; Japan	Global (low)	North Western Pacific, Japan regions, and Seto-Inland Sea (high)
NMEFC China	Global (medium)	North Western and Tropical Pacific, regional China Seas, and Indian Ocean (medium to high)
REMO Brazil	N/A	Atlantic (medium), Southern Atlantic and Brazilian Waters (high)
RTOFS NOAA; USA	Global (high)	North Atlantic and North Pacific (high)
TOPAZ NERSC; Norway	N/A	Arctic, North Atlantic (medium to high)

CHAPTER FOURTEEN

THE GODAE OCEANVIEW COASTAL OCEAN AND SHELF SEAS TASK TEAM

PIERRE DE MEY AND VILLY KOURAFALOU

Introduction

Starting in the mid-2000s, GODAE fully embraced the importance of coastal ocean "intermediate users" for coastal applications. To address this need, the Coastal and Shelf Seas Working Group (*CSSWG*, 2006–2008) was established with the mission to bridge global and coastal prediction strategies by quantifying elements that demonstrate the value of GODAE results for regional, coastal, and shelf seas models and forecasting systems. A total of 40 coastal ocean systems, each nested in larger-scale systems, in many coastal regions of the world ocean, were examined. Encouraging results from these and other systems were discussed at several events, especially regarding the suitability of the existing large-scale estimates, scientific quality, demonstrations of utility, as well as critical scientific issues and requirements related to downscaling (e.g., Alvera-Azcárate *et al.*, 2009; Counillon and Bertino, 2009; De Mey and Proctor, 2009; Halliwell *et al.*, 2009; Herzfeld, 2009; Kourafalou *et al.*, 2009; Le Hénaff *et al.*, 2009; Powell and Moore, 2009; Wakelin *et al.*, 2009).

With the evolution of the initial GODAE phase to GODAE OceanView (GOV) in 2008, the CSSWG was replaced by the Coastal Ocean and Shelf Seas Task Team (COSS-TT; https://www.godae-oceanview.org/science/task-teams/coastal-ocean-and-shelf-seas-tt/), which fully embraced the initial CSSWG recommendations, as detailed in De Mey *et al.* (2007). This is an international team, with members from every continent (specific affiliations are provided in the COSS-TT link). As GOV assigns itself broader objectives for the continuation of its global ocean-forecasting activities, COSS-TT has assumed the challenge of international coordination beyond the initial objectives of demonstration of feasibility and utility, to a next-level broader initiative aimed at consolidating the foundation and future advancement of coastal ocean-forecasting science, systems, and applications. The main goal and central mission of the COSS-TT is thus to work within GOV, and in coordination with GOOS, towards the *provision of a sound scientific basis for sustainable multidisciplinary downscaling and forecasting activities in the world's coastal oceans.* The initiative is built around three key concepts: "international", "scientific", and "sustainable", and is driven both by science and through the promotion of good practices, benefiting from overseeing the international coordination of a broad range of scientific phenomena and applications examined within individual Coastal Ocean Forecasting Systems (COFS).

The COSS-TT is one of the GOV Task Teams, but it has also initiated the consolidation of a broader coastal scientific community (so-called COSS-COMM) involved in developing and advancing methodologies supporting COFS. The COSS-TT decides targeted related actions at annual workshops that engage all TT members and the COSS-COMM. The strategic goal of the COSS-TT is to help achieve a truly seamless framework from the global to the coastal/littoral scale. The main disciplines considered by the TT are physics and interactions between physical and biogeochemical processes.

Links to Societal Benefits

Ocean forecasting has several key drivers related to the needs of national and international marine policies and management towards broad public, ecological, and economic benefits. These include: regulation of contracts and service agreements (environmental studies required for permits and oversight for marine construction, oil/gas exploration etc.); sea treaties and agreements (regulation of natural resources, marine pollution litigation, regional fish stock management, MPAs, United Nations Convention on the Law of the Sea, UNCLOS, etc.); and socioeconomic applications (recreational and commercial fishing, search and rescue, insurance claims, mitigation of natural disasters and extreme events, etc.). Although the COSS-TT does not coordinate these activities, it plays a central and relevant role, as per its mission of advancing science in support of coastal ocean forecasting. A major contribution of the COSS-TT is to address the particular challenges on monitoring and forecasting in coastal areas and regional seas, where the majority of human marine activities take place. As these are also the areas of enhanced exploitation of marine resources, the COSS-TT has a mission well aligned with society's needs and benefits.

The Task Team and its activities

COSS Task Team members are scientists, directly or closely associated with the development of COFS, who can (along with their teams) commit to community decisions established at TT workshops. These related actions, consolidated through detailed workshop reporting that is publicly disseminated, have broader implications, particularly for international agencies and organisations (such as the Regional Ocean Observing System/ROOS initiative).

The COSS-TT co-chairs are members of the GOV Science Team, representing the needs of the broader COSS-COMM in GOV. For instance, this interaction facilitates the availability of adequate large-scale ocean products for downscaling to coastal regions.

Specific TT activities include, but are not limited to:

- Reviewing progress with the COFS
- Establishing affiliated "geographic" groups (modelling the same region(s))
- Establishing international networks around common science themes (see next section)
- Organising science fora in dedicated sessions at international conferences (such as American Geophysical Union/AGU) following the same themes
- Providing feedback and recommendations for larger-scale GOV systems
- Providing feedback and recommendations for GOOS/ROOS and for Space Agencies
- Implementing a sustainable organisation for the TT itself.

TT members also serve as agents of scientific discussions and exchanges of innovative ideas and methodologies with the broader coastal community.

These activities are grouped into two major categories: (1) Establishing community links between on-going coastal ocean-forecasting projects; (2) Convening fora to discuss targeted science issues.

Establishing community links is achieved mainly through annual COSS-TT International Coordination Workshops (ICW). These aim to define and implement an adequate level of international scientific coordination among COFS, in order to advance science in support of coastal forecasting, but consider scientific as well as applied objectives. So far, two workshops (Miami, United States, 2012; Lecce, Italy, 2013) have been held, with increasing community participation and with comprehensive themes on all issues associated with the COFS needs and TT objectives (see below). A third workshop will be held in January 2014 in Puerto Rico. Each event comprises a Science Workshop open to the broader COSS-COMM and a restricted COSS-TT Meeting to prepare the following year's actions and Task Team roadmap. One important action is maintaining a Systems Information Table (SIT), based on a template, to be filled by each regional, shelf, or coastal system represented in the TT and Community. The ICW reports are publicly available on the COSS-TT web page.

Fostering links outside the community is also an on-going activity. The TT recently linked with the newly created EuroGOOS Coastal and Shelf Seas Modelling working group (Oddo and Leth, 2012), as well as with the coastal altimetry community (http://www.coastalaltimetry.org/). Benefits include the reciprocal participation of Coastal Altimetry (CA) and COSS-TT members in the respective annual workshops and the inclusion of TT outcomes in the reporting of CA activities (Cipollini *et al.*, 2012). In the second category of activities, sessions are regularly convened by the COSS-TT at AGU and Ocean Sciences meetings (every other year since 2008) along the themes detailed in the next section.

Science in support of coastal ocean forecasting

The following science themes have been found to be critical for quality downscaling and forecasting in the world's coastal oceans. They currently serve as the basis for the activities of the Task Team.

Coastal ocean science

The innovative approach that the COSS-TT advocates and oversees is that forecasting in the coastal and shelf seas must fully address land-sea, air-sea, and coastal-offshore interactions. It is important to stress that the COSS-TT and COSS-COMM have adopted the view that the influence of coastal ocean processes is felt far beyond the shelf break, thus interacting with open ocean dynamics and controlling the connectivity of remote ecosystems.

The coastal ocean processes cover mesoscale and sub-mesoscale shelf break exchanges, shelf dynamics, fronts, connectivity, slope currents, storm surges, tides, internal waves, surface waves, swell, upwellings, nutrients, sediments and pollutants, estuarine processes, plumes, and topographic controls on circulation. The land-sea interaction is governed by coastal runoff and the resulting buoyancy-driven circulation and material transport. The air-sea interaction processes are accommodated in the COFS through atmospheric forcing/coupling, allowing the appropriate modelling and impact of general circulation and extreme events. This requires *ad hoc* wind products and improved regional atmospheric forecasts via coupling with ocean models of appropriate resolution and dynamics.

Research, Development and Tools

Advancement of forecasting in coastal and shelf seas requires continuous development of innovative methodologies and tools, in support of research topics and applications. Coastal ocean models in COFS require dedicated numerical techniques for both individual ocean components and fully coupled models. In accordance with the COSS-TT strategic goal of a seamless framework from the global to the coastal/littoral scale, issues of downscaling from larger scale models and appropriate nesting procedures have to be addressed (see Figure 14-1). Assessment of the boundary conditions provided to COFS nested systems are an important link with the GOV national systems, aiming at a dynamic dialogue that will help improve both global and coastal/regional forecasts. Other primary topics driving COFS research and development include: data assimilation challenges that are unique to coastal ocean dynamics and available observations (see next section), and assessment of model errors (ensemble/stochastic approaches).

Observation-related issues

Observations are crucial for model evaluation and assimilation. The COSS-TT focuses on issues unique to coastal ocean observing systems, in support of advancing the accuracy of COFS forecasts. In addition to satellite observations covering the global oceans, ship surveys, and global standards for drifting and moored instrumentation, the coastal ocean monitoring systems include dedicated or suitably modified instruments. Examples include coastal radar, shallow water drifters, and coastal altimetry products. International coordination through COSS-TT assures exchange of experience on best practices for model validation strategies and metrics. An important activity is the evaluation of observation impact on COFS forecasts through rigorous methodologies that quantify the performance of coastal observational arrays and provide assessment of their errors. OSSEs provide a rigorous, cost-effective approach to evaluate the impact of new observing systems and alternate deployments of existing systems, and to optimise observing strategies for field experiments. OSSEs are an extension of OSEs which use data denial experiments to determine the impact of existing observing systems. These techniques, adapted for the coastal ocean, are important tools for the optimisation of existing, and the planning of future coastal observing systems (e.g., Surface Water Ocean Topography (SWOT) satellite mission).

Application-targeted science

COSS-TT supports application-targeted science studies that focus on linking applications with underlying predictable variables in COFS, and on determining desirable prediction ranges and accuracy. These include:

- Sea level changes: e.g., storm surges and flooding
- Surface currents, near-shore currents: e.g., maritime safety, oil spills, radionuclide dispersals, and environmental protection
- Sediment dynamics and transport; turbidity
- Temperature, density, and vorticity: e.g., fisheries, aquaculture and sonar prediction
- Water quality and the fate of pollutants
- Coastal ecosystem impacts and responses.

As an example of the latter category, we are in the process of linking Phenomena of Interest (PoIs) defined by the GOOS Panel for Integrated Coastal Observations (PICO) Plan (see Chapter 16) with predictable variables in the COSS-TT COFS. An example of data harvested from the COSS-TT Systems Information Table (SIT) is shown in Table 14-1.

Conclusion

The Coastal Oceans and Shelf Seas Task Team has set for itself the objective of advancing science in support of sustainable multidisciplinary downscaling and forecasting activities in the world's coastal oceans and their applications, in connection with the larger-scale ocean forecasting activities in GOV, and with observation-related activities taking place in

GOOS and ROOS. These objectives are pursued through year-round international community activities and fine-tuned as needed at annual workshops. The first outcome of these workshops is to provide international networking among the traditionally fragmented coastal ocean modelling community, the growing list of operational coastal ocean forecasting systems engineers, the larger-scale providers, the observation specialists, and people and companies interested in applications. The second major outcome is to provide a clearer picture of the status of operational coastal oceanography around the world; the COSS-TT maintains a Systems Information Table to that end. A third outcome is the series of fora in international conferences and workshops for cross-fertilization of ideas and methodologies through dedicated sessions on themes relevant to both science and applications, including integrated observations, downscaling approaches, coastal model validation approaches, and data assimilation. A fourth outcome is to make GOV quality-controlled, large-scale data providers visible to COSS modellers, and to relay the COSS needs to these large-scale providers.

Main challenges in fulfilling the TT's specific objectives and overarching goals lie in the multiple interests and requirements of individual groups composing the COSS Community, which is actually still emerging. One action that the TT is pursuing to help consolidate the Community is to facilitate the launching of Pilot Projects among two or more international members and regional groups. In that spirit, the COSS-TT is currently preparing the 3rd international coordination workshop in Puerto Rico, in early 2014. Furthermore, a Memorandum of Understanding (MOU) has been crafted during the 2nd workshop to ensure institutional, long-term support to TT activities by all member-signatories. The MOU is currently being revised to accommodate a broad agreement by all TT member institutes, and will be presented for final approval in the upcoming 3rd workshop.

Overall, the COSS-TT is dedicated to its vision of a strong, diverse Coastal Ocean and Shelf Seas Community that will carry science-based and technologically advanced coastal ocean forecasting into the future and support a broad range of applications to benefit society.

Table 14-1: PICO Phenomena of Interest (PoI), corresponding indicators, and occurrence of these PoIs in the Objectives section of the Systems Information Table (SIT), on a total of 31 answers. The total can be larger than 31 since several choices were allowed

PoI	Key Indicators of Ecosystem States	Response (SIT)
Harmful Algal Blooms	Distribution and abundance of toxic phytoplankton species	11/31
Coastal Eutrophication & Hypoxia	Phytoplankton biomass fields Dissolved oxygen fields	7/31
Vulnerability to coastal flooding	Extent and condition of habitat buffers to flooding	7/31
Food Security	Abundance of harvestable finfish and shellfish stocks	5/31
Ocean acidification	Extent and condition of coral reefs Abundance of calcareous plankton	4/31
Habitat Loss & Modification	Extent and condition of biologically structured habitats	3/31
Human Exposure to Waterborne Pathogens	Distribution and abundance of waterborne pathogens	1/31

OPERATIONAL OCEANOGRAPHY IN BRAZIL: A CONTRIBUTION TO MONITORING AND PREDICTING THE TROPICAL AND SOUTH ATLANTIC

CLEMENTE A.S. TANAJURA, PAULO NOBRE AND EDMO J.D. CAMPOS

Introduction

This chapter presents an overview of the main operational oceanography activities in Brazil. It addresses observation, modelling, and forecasting, with a focus on the tropical and South Atlantic. It also presents details of three activities to better disseminate information from Brazilian efforts in this area.

International cooperation

Operational oceanography in Brazil has been very much associated with *in situ* observations, and the Brazilian Navy has occupied a central and historical role. Hydrography and operational oceanography in Brazilian waters have been mostly conducted by the Brazilian Navy Directorate of Hydrography and Navigation (DHN) and its Navy Hydrographic Centre (CHM).

In the last three decades, joint collaborations among the Brazilian Navy; the Brazilian Ministry of Science, Technology, and Innovation (MCTI), particularly through its National Institute for Space Research (INPE); and universities, mainly the University of São Paulo (USP) and the Federal University of Rio Grande (FURG), have been very successful. They have led to important participation of Brazil in on-going national and international observational programmes, such as the Prediction and Research Moored Array in the Tropical Atlantic (PIRATA), the National

Buoy Program (PNBOIA), the Argo programme, the Global Sea Level Observing System (GLOSS), the Antares network for coastal ecosystem studies, and the programme on Monitoring the Upper Ocean Thermal Variability between Rio de Janeiro and Trindade Island (MOVAR).

These programmes were organised in Brazil under GOOS-Brasil (http://www.goosbrasil.org), the Brazilian national ocean observing system. It aims to collect and distribute quality-controlled ocean data in the tropical and South Atlantic in operational mode for monitoring and research purposes. In addition to the PIRATA moorings, the system encompasses: six moored buoys along the Brazilian shelf with sensors for air and water temperature, winds, and currents, among other met-ocean variables under PNBOIA; 12 tide gauges under GLOSS; and, since 2004, a high density XBT line (NOAA AX97) with temperature measurements in the top 800 metres roughly every two months under MOVAR.

Another international programme in which Brazil plays a major role is the India-Brazil-South Africa Ocean (IBSA-Ocean) programme. It focuses on areas relevant to operational oceanography over the Tropical and South Atlantic, namely: (i) climate variability and climate change, and (ii) regional ocean observing systems. This programme has brought a different twist to the operational oceanography in the Atlantic, as it brings a broader spectrum of activities from sensor and observing systems development to international joint research and capacity building in the region. The first IBSA-Ocean summer school on global climate modelling was conducted in Brazil in late 2011, with the participation of doctoral and post-doctoral fellows from all three IBSA countries, along with other South American countries.

Nationally, Brazil is developing a major structured research and operational framework on oceanography, with the recent creation of its National Institute for Ocean Research and Waterways (INPOH); the acquisition of trans-Atlantic oceanographic research vessels, both abroad and in the country; and investments in the expansion of its ocean monitoring array. As a result of a partnership with the NOAA/Pacific Marine Environmental Laboratory (NOAA/PMEL), Brazil has just completed the construction and deployment of its first moored system for deep waters, the ATLAS-B system. The first of these buoys was deployed in April 22, 2013 near 28.5°S, 44°W, a region where Catarina – the first ever documented hurricane in the South Atlantic – formed in 2004. Repeat oceanographic cruises to serve this buoy will complement the *in situ* data. This will allow the assessment of seasonal and longer-term variability, thus contributing to the understanding of how climate change can affect physical, chemical, and biological processes in the region.

Regarding operational ocean short-range forecasts and ocean data assimilation, a consortium was formed among DHN/CHM, a number of universities and the Brazilian state oil company Petrobras: the Oceanographic Modelling and Observation Network (REMO) (http://www.rederemo.org). Its goals are: (i) to undertake research and technological development to support operational oceanography, considering modelling, observation, data assimilation, and forecast, and (ii) to provide oceanographic information on a daily basis for several purposes, including oil exploitation, navigation safety, environmental licensing, and military applications (Lima *et al.* 2013). REMO has been a member of the GODAE OceanView project (see Chapter 13) since the end of 2010, which in turn has helped REMO to develop quickly. In 2013, REMO has also become a member of GHRSST, an international group dedicated to the development of high resolution SST products for the scientific community.

On the climate modelling side, Brazil is constructing its Brazilian Earth System Model, which presently is a fully-coupled global ocean-atmosphere climate model with marine ice (Nobre *et al.*, 2012; 2013) and marine biogeochemistry (Farias *et al.*, 2013). In the future, it shall incorporate a dynamical vegetation component model and atmospheric chemistry model. Such developments are the result of a concerted set of actions of the Federal Government of Brazil and State Governments, through the creation of the National Network on Global Climate Change Research (Rede CLIMA), the National Institute for Science and Technology on Climate Change (INCT-MC), and the State of São Paulo Research Foundation (FAPESP) Research Program on Global Climate Change (PFPMCG), among other Federal initiatives.

Below, details of three Brazilian operational oceanography initiatives are highlighted: PIRATA, the recently established South Atlantic Meridional Overturning Circulation Programme (SAMOC) and REMO.

The Prediction and Research Moored Array in the Tropical Atlantic (PIRATA)

PIRATA is a trilateral partnership between Brazil, France, and the United States to create and maintain an *in situ* operational observation system in the tropical Atlantic (Bourlès *et al.*, 2008). The main goals of PIRATA are: (i) to improve the description of the intraseasonal-to-inter-annual climate variability in the tropical Atlantic; (ii) to support the investigation of physical mechanisms associated with climate variability;

and (iii) to provide data for model development, data assimilation and predictability improvement in short and longer timescales.

PIRATA started as the Pilot Research Moored Array in the Tropical Atlantic in 1997 with the implementation of 6 Autonomous Temperature Line Acquisition System (ATLAS) moored buoys in the eastern and in the western tropical Atlantic. The buoys were equipped with atmosphere and ocean sensors to measure surface heat and moisture fluxes, SST and salinity sensors, and subsurface temperature and salinity sensors in the upper 500 m. The backbone of the system, composed of 10 buoys, was fully implemented in 2001. However, because of the desire of the participating countries to continue the project, expand the array, and transform it into a permanent array, extensions were implemented after 2001. These extensions were named Southeast Extension, Southwest Extension, and Northeast Extension. Today, a total of 18 met-oceanographic buoys and one moored ADCP are operational (see Figure 15-1), and some ATLAS moorings also measure carbon dioxide air-sea flux. Therefore, the PIRATA backbone provided the basis not only for an increase in the number of moored buoys, but also enabled the construction of a more sophisticated monitoring system.

The PIRATA array is complemented by three island-based observation sites, two in the western equatorial Atlantic and one in the eastern equatorial Atlantic. Brazil (the Brazilian Navy and MCTI), is responsible for retrieving and re-deploying the buoys in the western part of the array, while France and the United States are responsible for the eastern side. During each cruise for maintenance of the array, oceanographic vessels also collect complementary data, including surface and upper air meteorological variables, surface and sub-surface currents, and biogeochemical data that are employed in the validation of ocean and/or coupled ocean-atmosphere models.

Data from the entire array are transmitted hourly to operational oceanographic and meteorological centres around the world via the GTS. Over 235,000 data files were delivered via the internet during 2012 alone; and with 15-plus year time series of daily buoy data available for research, the PIRATA array constitutes, *de facto,* the fixed buoy reference array over the Tropical Atlantic, recognised by the CLIVAR and OOPC. Additionally, Brazil has also developed an ATLAS like mooring system, the ATLAS-B system, which is based on NOAA/PMEL's ATLAS, but with a new electronics and off-the-shelf set of sensors. The first ATLAS-B system was moored in deep waters off the coast of Santa Catarina in southern Brazil for test purposes, and functioned uninterruptedly for a test period of six months.

The South Atlantic Meridional Overturning Circulation Programme

SAMOC is a collaborative effort, involving investigators from France, Brazil, United States, South Africa, Argentina, Russia, and Germany, to monitor the Meridional Overturning Circulation (MOC) in the South Atlantic (Schiemeier, 2013). It is endorsed by the project CLIVAR (Climate Variability and Predictability) of the World Climate Research Programme. France, Brazil, Germany, and the United States will provide the major instrumentation for a moored array along 34.5°S, the backbone of the SAMOC field programme named the SAMOC Basin Wide Array (SAMBA). South Africa, Russia, Brazil, Germany, and Argentina will contribute funds for ship-time and local-expertise for the turn-around and recovery cruises. Brazil, via the Oceanographic Institute of the University of Sao Paulo, in cooperation with Argentina, and the United States NOAA Atlantic Oceanographic and Meteorological Laboratory (NOAA/AOML), will conduct repeat hydrographic cruises and maintain an array of current meters and other sensors at the western end of the SAMBA line. France and Brazil will lead the development of a common strategy in regional climate models to downscale climate variability and assess the ocean circulation's influence on climate change and its impact over South America and Africa.

The Oceanographic Modelling and Observation Network (REMO)

Since 2007, a specific effort in operational oceanography with focus on ocean weather forecasts with high-resolution models, ocean data assimilation, and observations in the tropical and South Atlantic, was initiated with the organisation and implementation of REMO. It is a consortium formed by the Federal University of Bahia, Federal University of Rio de Janeiro, University of São Paulo, the Brazilian Navy Hydrographic Centre (CHM) and the Research and Development Centre of the Brazilian-state oil company Petrobras (CENPES). REMO is sponsored mainly by Petrobras and the Brazilian National Agency for Oil, Natural Gas and Biofuels (ANP) (Lima *et al.*, 2013).

REMO's products complement operational ocean wave and marine weather predictions performed by the CPTEC/INPE and the Brazilian Navy CHM, and the weather and climate forecasts produced by CPTEC/INPE and the National Institute of Meteorology (INMET). The main regions of interest are the Metarea V – Atlantic waters west of 20°W from approximately 35°S to 7°N – and the whole Brazilian continental shelf, since there is high demand for oceanographic information and forecasts by Brazilian authorities in these regions. For instance, the Brazilian Navy is responsible for the dissemination of met-ocean information and safety of navigation in Metarea V, including search and rescue missions. Also, 90% of the oil and gas consumed in Brazil is produced offshore along the Brazilian continental shelf.

Regarding observations, cloud-free SST composition method was developed to construct daily SST analyses with 0.05° resolution from satellite data. The analyses are validated against *in situ* data. REMO financed the purchase of one moored buoy and it contributes with funds to the system maintenance. Two pairs of moorings will also be deployed off Cabo de São Tome and Cabo Frio on the southeast Brazilian shelf, critical regions where the variability of the Brazil Current is high.

Regarding modelling and forecasting, a nested operational prediction system based on the Hybrid-Coordinate Ocean Model (HYCOM) (Bleck, 2006) was implemented at CHM in February 2010 and upgraded in February 2011 (Lima *et al.*, 2013). The Regional Ocean Modelling System (ROMS) is also being used in REMO (Marta-Almeida *et al.*, 2011; Amorim *et al.*, 2012) and should soon become operational. The entire Brazilian shelf will be covered by HYCOM and/or ROMS configured with horizontal resolutions ranging from 1/24° to 1/36°.

Every day, REMO's operational system produces 3-to-7-day ocean forecasts forced by atmospheric forecasts from the regional model HRM run at CHM and/or from the Global Forecast System of the United States National Centres for Environmental Predictions. The forecasts and the SST analyses are routinely made available via the portal at http://www.rederemo.org and via the CHM portal as part of the Marine Meteorological Service https://www.mar.mil.br/dhn/chm/meteo/prev/mod elos/hycom-v.htm.

Links to Societal Benefits

The Brazilian efforts to develop operational oceanography in the tropical and South Atlantic meet the interests of several societal sectors in Brazil and abroad. The Brazilian oceanic exclusive economic zone is vast, and together with the area surrounding the Brazilian Trindade Island, it adds up to almost 4 million km^2. In Brazil, this large area has been called the Blue Amazon, since it is about the size of the Brazilian Amazon rainforest. Surveillance and environment sustainability of the Blue Amazon is therefore a major challenge, particularly due to the important oil and fishery industries established in this region. Monitoring and forecasting the ocean circulation and physical state is crucial to safe operation of these industries, to environmental control and to a fast response to environmentally threatening accidents.

As mentioned above, Brazil is responsible for search and rescue missions in Metarea V, which is much larger than the Blue Amazon. PIRATA observations, REMO forecasts and remotely sensed data provide important information and scenarios in near-real time to Brazilian authorities, particularly to the Brazilian Navy, to narrow down the search and rescue area. Other sectors like commercial navigation and tourism also benefit from the products made available by the aforementioned Brazilian programmes, but efforts to enhance the engagement of the users of the oceanographic information are still necessary.

The links between operational oceanography and climate open another dimension of societal benefits produced by the Brazilian programmes. For instance, the South Atlantic transports a huge amount of heat to the North Atlantic and it is a key component of the global MOC, an important climate regulator. There is observational and model evidence that the South Atlantic salinity increased in the past decades and that increased CO_2 concentration may further intensify this signal (Biastoch and Böning, 2013) and consequently modify the MOC. There is no doubt that the Brazilian operational oceanography is contributing to a better understanding of the present climate variability and climate change scenarios.

Conclusion

Brazil is quickly becoming a major international player in operational oceanography for the tropical and South Atlantic. Both at home and abroad, the country is taking solid steps towards advancing ocean research and operational oceanography as a major component of national development.

Acknowledgments

In Brazil, SAMOC is funded by FAPESP (Proc. 2011/50552-4). Atlas-B is an activity of the INCT for Global Change. The first author would like to acknowledge support of the CAPES Foundation, Ministry of Education of Brazil, (Proc. BEX 3957/13-6).

PART V

SERVICES FOR THE COASTAL ZONE

GLOBAL OCEAN OBSERVING SYSTEM (GOOS) REGIONAL ALLIANCES, PANEL FOR INTEGRATED COASTAL OCEAN OBSERVATIONS (PICO): REQUIREMENTS FOR GLOBAL IMPLEMENTATION OF THE STRATEGIC PLAN FOR COASTAL GOOS

ZDENKA WILLIS, LAURA GRIESBAUER, PAUL DiGIACOMO AND JOSE MUELBERT

Introduction

The Global Ocean Observing System (GOOS) is a permanent global system involving 105 nations for observations, modelling, and analysis of marine and ocean variables to support operational ocean services worldwide. GOOS provides accurate descriptions of the present state of the oceans, including living resources, continuous forecasts of the future conditions of the sea for as far ahead as possible, and the basis for forecasts of climate change. GOOS is sponsored by the International Oceanographic Commission (IOC), the United Nations Environment Programme (UNEP), the WMO and the International Council for Science (ICSU), and is the oceanographic component of GEOSS. GOOS is implemented by member states via their government agencies, navies and oceanographic research institutions working together in a wide range of thematic panels and regional alliances.

Elements of international collaboration

GOOS Regional Alliances (GRAs) are tasked to bring the GOOS principles of shared ocean observations, data policy, best practices and capacity development to regional and national ocean observation systems. Historically the GRAs were introduced as a way to integrate national needs into a regional system with bi- and multi-lateral coordination between adjacent states. The first GRAs formed in 1994 and 1996. Since that time, the GRAs have evolved to meet the needs of society in a wide range of societal benefit areas through both coastal and open-ocean observing.

GRAs are comprised of national efforts that come together at the regional scale to facilitate the advancement of the GOOS and to aid the integration and coordination of sustained interdisciplinary ocean observations for scientific and societal benefits. Although GRAs are not distinctly globally or coastally focused, responding instead to the needs of the national efforts they represent; they are well suited to accelerate the integration and expansion of observations in the coastal zone globally.

GRAs should strive to:

- Serve as a platform for the coordination of sustained regional observations, transboundary observing networks, and a link to GOOS/JCOMM; provide access to real-time and archived data streams, from *in situ* and relevant satellite observations; create information products and model outputs; conduct assessments of regional readiness and capacity in each of the areas above and asses the overall performance of the system in providing users with fit-for-purpose data and information products.
- Promote programmes on developing regional capacity through the sharing of experience, success stories, and best practices.
- Develop the ocean observing community through scholarships, exchanges, technical skills workshops, and programmes to develop leadership and grant-writing skills.
- Encourage the development of Regional and National Ocean Observing Systems by promoting the visibility and recognition of the services provided by ocean observing systems with governmental agencies and private companies; encourage integration at national, regional and global levels, advancing the scientific and technological developments upon which services

depend; and identify gaps in ocean observations at regional and national levels.

While every GRA is unique, a review of the assessments every GRA completed in 2013 showed some common elements:

- The majority of the GRAs come together as an association with various governance structures of steering teams; two are national programmes, and one has created a non-profit organisation.
- Common themes include: coordination of observing systems across member nations, contribution to and advancement of GOOS, exchange and access of oceanographic data, development of downstream services for end-users and strengthening of capacity building.
- Readiness of the observation network and data management varies from 0-100%, but the rationale provided gives us a concrete plan on how to advance the capability.
- Modelling capability focuses on physical models with several pilot projects that include the coupling of social and ecosystem modelling with physical modelling.

Links to Societal Benefits

The GOOS Panel for Integrated Coastal Ocean Observations (PICO) delivered GOOS Report #193, *Requirements for Global Implementation of the Strategic Plan for Coastal GOOS*, in 2012. While referred to as "Coastal GOOS", the scope of the PICO effort is not only estuaries and the coastal ocean, but also the ocean basins as well. The goal of this report was to develop a plan for expanding GOOS for the sustained provision of data and information to inform Ecosystem Based Approaches (EBAs), for the management of human uses of ecosystem goods and services, and for adaptation to climate change on local to global scales.

In this context, the PICO report builds on analyses and recommendations of the earlier GOOS Coastal Ocean Observing Panel (COOP), the Coastal Theme of the Integrated Global Observing Strategy (IGOS), OceanObs'09, *A Framework for Ocean Observing*, and *An Assessment of Assessments* of the United Nations, and provides a plan for expanding GOOS to include biogeochemical and ecological elements for coastal waters and the ocean basins. Its recommendations are intended to complement and leverage those aspects of the ocean-climate system addressed by the OOPC and existing operational programmes for predicting extreme weather events and tsunami, changes in physical states of the upper ocean, and coastal flooding.

An end-to-end observing system approach was highlighted within the PICO Report. The PICO report identified the following seven priority Phenomena of Interest:

- Coastal Eutrophication and Hypoxia
- Human Exposure to Waterborne Pathogens
- Harmful Algal Blooms
- Habitat Loss and Modification
- Vulnerability to Coastal Flooding
- Ocean Acidification
- Food Security.

For each Phenomenon of Interest the following were identified:

- Primary user groups and their data and information requirements for products and applications;
- Key indicators of relevant pressures, states and impacts of changes in state;
- Observing system requirements, spanning observations (*in situ* and remote sensing), modelling and analysis, reporting (real–time or delayed mode), and data management and communications; and,
- Operational status for each attendant observing system and gap analysis to help determine priorities for capacity building and research.

Relative to the phenomena of interest, the following priority indicators of ecosystem states (ecosystem health) were identified as targets for specifying observing system requirements:

- Phytoplankton biomass and oxygen fields;
- Distribution and abundance of waterborne pathogens;
- Distribution and abundance of toxic phytoplankton;
- Spatial extent of benthic biological habitats;
- Ecological buffers to coastal flooding;
- Distribution and condition of calcareous organisms; and
- Distribution and abundance of exploitable fish stocks.

Based on the data requirements for each end-to-end solution, a suite of associated essential ecosystem state variables (geophysical, chemical, biological and biophysical) and essential pressure variables were in turn identified.

The following is an end-to-end solution for indicators of habitat loss and modification, specifically highlighting coral reefs, sea grass beds, mangrove forests and salt marshes. This example was selected because rising sea level, extreme weather, and urban development are expected to triple the number of people exposed to coastal flooding by 2070. These habitats are critical to the provision of goods and services including:

- Supporting high species diversity and living marine resources;
- Buffering coastal communities and ecosystems against storm surges and flooding;
- Functioning as carbon sinks (mangrove forests in particular);
- Providing important indicators of the impacts of ocean warming and acidification (coral reefs in particular);
- Tourist attractions.

The objective of the GOOS RA is to build a system that will monitor the changes and fragmentation of habitats that buffer coastal ecosystems and human populations against rising sea level and coastal flooding; provide assessments of current vulnerability to flooding; and develop realistic scenarios for how changes in habitat buffers will impact the vulnerability of future populations to coastal flooding in the long term and provide forecasts of the risk of exposure to waterborne contaminants associated with post-flooding runoff events in the short term.

Products of this end-to-end observing system include digital maps of vulnerability levels updated every 1–5 years; scenarios of future vulnerability; and nowcasts and forecasts of water quality parameters updated daily until the event signature dissipates. Observation requirements include data from both *in situ* and from remote sensing assets and reporting in near real-time and delayed mode. Essential variables include sea level, rain fall, river flows, water temperature and salinity, surface currents, and wave fields. Requirements from models include high resolution digital elevation models of topography, shoreline position and bathymetry; algorithms to compute vulnerability levels; hydrodynamic models; and maps of levels of vulnerability and post-event temperature, salinity, chlorophyll, dissolved oxygen, suspended sediment and contaminants.

A number of implementation priorities were identified to advance the overall end-to-end observing system capabilities recommended by PICO:

- Support national and international programmes that target priority infrastructure for key observations and predictions as identified in the PICO report.
- Establish data management and communications systems for interoperability among monitoring systems and data integration within and among regions.
- Support capacity building and research and development to fill priority spatial and temporal gaps in the global coastal network.
- Facilitate regional implementation of a pilot project in a priority "super site" domain to demonstrate the value added of an end-to-end system of systems (e.g., multiple applications of data and information needed to guide EBAs derived from a common set of observations and models).

Addressing these priorities will require investments by developed nations to ensure the coordinated establishment of a global network of national and regional observing systems that are locally relevant and interoperable in terms of data and information exchange. Such mechanisms must:

- Engage groups that use, depend upon, manage, and study marine systems in the design, operation and evolution of a coastal GOOS that meets their data and information needs on local to global scales;
- Build on and leverage existing programmes with common goals and objectives;
- Promote the development of regional observing systems and services in regions populated by developing countries;
- Promote the development of a Global Coastal Network (GCN) through coordinated regional development worldwide; and,
- Effectively interface with the existing planning, oversight and implementation bodies of GEOSS, GOOS, GCOS, the Global Terrestrial Observing System (GTOS), and other organisations as appropriate.

Conclusion

Successful implementation of coastal observations depends on more effective collaboration with stakeholders across the land-sea interface, such as the GEOSS CZCP (see also Chapter 17). The CZCP supports the operational goal of providing data and information needed to inform management decisions across the land-sea interface. The PICO plan can be used as a blueprint for the GRAs and CZCP to build out the coastal component of GOOS based on pressing issues facing the large proportion (more than 50%) of the world's population that lives within 50 miles of the ocean.

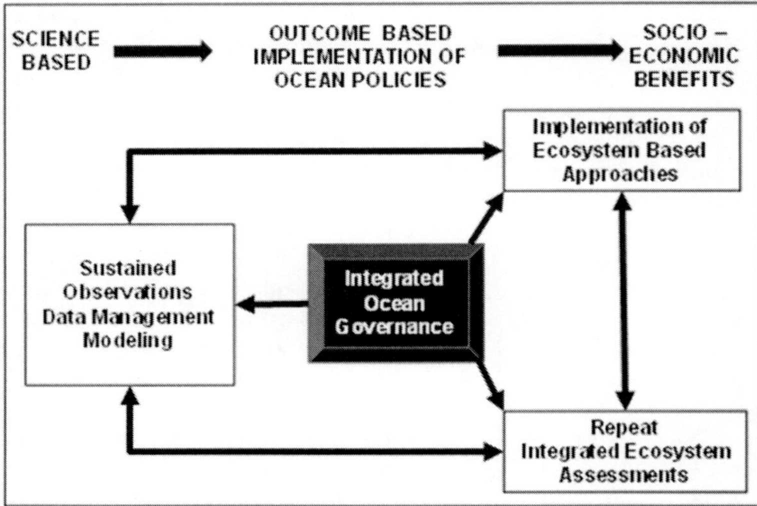

Figure 16-1: Implementation of scientifically sound ocean policies for sustained development (socio-economic benefits) depends on a closely coupled system of integrated and sustained ocean observations (GOOS), repeat computation of indicators and assessments, and implementation of ecosystem-based approaches. Given that ecosystems are complex systems characterised by many interacting properties and processes that cannot all be monitored in all places at all times, it is important to identify key ecological indicators to implement performance-based ocean policies and the ecosystem-based approaches called for in these policies. IOC Document 193

CHAPTER SEVENTEEN

THE COASTAL ZONE COMMUNITY OF PRACTICE:
SUPPORTING INTEGRATED COASTAL ZONE
MANAGEMENT WITH EARTH OBSERVATIONS

MILTON KAMPEL, PAUL DIGIACOMO
AND HANS-PETER PLAG

Introduction

Coastal zones are complex dynamic areas of significant ecological, social, and economic value where many conflicting interests need to be balanced in order to ensure sustainable development. Growing coastal population, increasing urbanisation and climate change impacts are elevating environmental stress in coastal zones, posing serious risks to human health and safety as well as to economic productivity and reducing the capacity of coastal ecosystems to support the provision of critical goods and services.

Increases in coastal populations and urbanisation and changes in land-use practices in coastal catchments and floodplains have led to rapid and large changes in sediment supplies, and increases in nutrient, pollutant, and pathogen loadings to coastal waters, as well as greater susceptibility to episodic and sustained flooding. These growing pressures pose serious risks to humans and coastal ecosystems as a whole, in developed and developing nations alike. Risks associated both with slow changes in the means and extreme events are increasing and are very likely to be compounded by global climate change. In this context, high priorities for GEOSS include improved forecasts of local sea-level rise, the associated increases in coastal inundation and the changing probability of storm surges. The latter may be exacerbated by increases in the frequency of extreme weather events. Likewise monitoring and forecasting of water quality and associated states of ecosystem health and productivity are supported by GEOSS.

Detecting, assessing, predicting and managing the interplay of coastal urbanisation and climate change is critical to the sustainability of healthy and resilient coastal ecosystems and the products and services they provide to human populations. Thriving coastal communities require comprehensive information on the state and trends in the environmental, social, and economic conditions of the coastal zone. Approaches for performing these tasks will necessarily differ from country to country in view of the multiplicity of tools, policies, measures, and standards employed for assessing and managing risk within coastal zones. Nonetheless, there remains an urgent need to promote integrated, multi-disciplinary (including ecosystem-based approaches), and multi-sectoral coastal and ocean governance at the regional level in ways that leverage, support, and enhance the capacity of individual coastal states in developing effective coastal zone policies, robust response mechanisms, and risk mitigation strategies.

Improved, integrated and sustained coastal observing capabilities are required to better support user information needs. GEO provides a valuable framework and mechanism to help implement these capabilities and, under the auspices of GEO, the Coastal Zone Community of Practice (CZCP) was initiated in 2006.

The CZCP supports GEO in its goal to provide timely observations to inform decisions related to the coastal zone. Focus is not only on the provision of observations but also the delivery of practice-relevant information. High priorities for coastal users of GEOSS include improved forecasts of local sea-level rise on inter-annual to decadal time scales as a basis to assess associated increase in frequency and extent of coastal inundation (Plag *et al.*, 2010), which may be further exacerbated by an increase in the frequency of extreme weather events. Assessing changes in water quality and broader ecosystem health, productivity and ability to provide important goods and services are also focus areas of the CZCP, particularly for mangroves, coral reefs and estuaries.

Through provision and utilisation of Earth observations and derived information, the CZCP brings together experts from multiple disciplines to support integrated and adaptive approaches for sustainable development in the coastal zone. The CZCP focuses both on research and practice-relevant knowledge applications related to coastal zone management and deliberative governance (Glavovic, 2013) in the coastal zone. It interacts with GEO's Societal Benefits Implementation Board (SBIB) and draws on support from various GEO Participating Organisations and related entities.

The CZCP emerged from the Coastal Theme of the Integrated Global Observing Strategy Partnership (IGOS-P), published in January 2006 by the IOC. The report is available at http://www.czcp.org/library/reports.php (see also a related article on coastal observing systems by Christian *et al.*, 2006). A recent report from the GOOS Panel for Integrated Coastal Observations (PICO) addressed the global implementation of the coastal component of GOOS (PICO-I, 2012), and identified a number of important functions that can be fulfilled by the CZCP, including observing system gap analyses, user need assessments (e.g., Plag *et al.*, 2012), as well as technical oversight.

GEO CZCP objectives and activities

The CZCP is a user-driven community of stakeholders whose purpose is to develop and foster, within the framework of GEO, a strategy for engaging and supporting coastal users across the land-sea interface. It achieves this goal by facilitating the development, implementation and integration of those elements of GOOS, GTOS and GCOS that provide timely, accurate, and fit for purpose data, products and information on terrestrial, aquatic (freshwater and marine), and atmospheric processes and phenomena and their convergence in the coastal zone, relevant in the context of prevailing and emerging socio-economic and environmental issues, needs and concerns.

The primary CZCP objectives are to:

- Engage coastal users (e.g., managers and decision makers) and data providers to collaborate in the collection of user needs in terms of information and, based on that, to derive specification of requirements for *in situ* and remote coastal observations and derived information products. The CZCP utilises for this the GEOSS User Requirements Registry (URR) (see Plag *et al.*, 2012) which ensures the integration of this information into an interdisciplinary knowledge base of user needs and observational requirements. These requirements identify which variables to monitor, their temporal and spatial resolution, accuracy, frequency, latency and delivery methods;
- Evaluate current and projected observational capabilities against these requirements, identifying gaps, redundancies, and activities that need to be strengthened and implemented;
- Organise workshops and promote "proof of concept" pilot projects that both address these gaps and enable data integration for the provision of new or improved decision support tools;

- Engage in the development of a comprehensive Global Coastal Zone Information System (GCZIS) as a platform for co-design and co-creation of knowledge with coastal stakeholders;
- Promote the development and strengthening of networks of institutions – globally, regionally, and across Communities of Practice (CoPs) – that contribute to, and benefit from, GEOSS;
- Advise the SBIB, other CoPs, and GEO, on matters relating to coastal zone observations and related societal benefits, and identify priority needs as well as areas for linkage, collaboration, and support.

CZCP coordination and support activities to date have included the organisation of a series of regional workshops (Mediterranean, Africa and the Caribbean); development of a web site that currently provides access to information about collaborative community activities and resources, and will ultimately provide key coastal data, products and information; identification of observing system needs and gaps; facilitating data identification, utilisation and integration for the provision of new or improved decision support tools, and support for diverse coastal applications and services.

Other general CZCP activities include:

- Building on the legacy of the IGOS Coastal Theme Report, bridging data providers and users;
- Populating the URR;
- Interfacing with GOOS, GTOS, GCOS and other global and regional observing systems and helping to inform and implement dedicated coastal observing capabilities;
- Facilitating data sharing, access and integration;
- Building and sustaining capacity;
- Identifying resources and collaborative partnerships to develop applications and services.

As above, CZCP also aims to support and facilitate coastal user related pilot projects and technical activities, as well as general user outreach and engagement activities, such as meetings, workshops and training programmes. However, the CZCP is currently constrained in its scope and efficacy, as it has minimal resources and operates on a best-effort basis, all of which limit the service it can provide in support of coastal users across the globe.

CZCP engagement with GEO Blue Planet

CZCP is also identified in the 2012-2015 GEO Work Plan as providing integral support to the implementation of the "Oceans and Society: Blue Planet" Task, specifically the implementation of a new component: *Services for the Coastal Zone*. Under this component, a suite of priority actions and activities have been identified:

- Develop a global coastal zone information system (GCZIS): a global cyber-infrastructure that will provide access to available information on coastal zones and facilitate the collection of new information through crowd-sourcing and citizen-science;
- Implement a pilot project in a priority, at-risk region (e.g., Indonesian Archipelago-South China Sea domain) to demonstrate the added-value of ecosystem-based approaches for monitoring and managing the coastal zone, engaging with and coordinating across GRAs and other global and regional ocean observing networks as discussed in the PICO plan for implementation of coastal GOOS (PICO-I, 2012);
- Assess climate change impact on the coasts of the Caribbean islands as a demonstrator for the application of space-based observations, fully engaging with CEOS with regard to data needs;
- Assess user needs and observational requirements for coastal water quality; identify indicators and best practices for coastal water quality, and implement a monitoring service pilot for coastal water quality; disseminate information particularly to under-served communities; and
- Assess the observational requirements for decadal forecasts of coastal local sea-level variation and develop a demonstrator forecasting service.

The GCZIS will be a key deliverable and outcome, which will be implemented as a globally available cyber-infrastructure that provides access to timely, accurate and fit for purpose data, products and information, utilising observations from global providers (e.g., space agencies) as well as contributions from citizen-scientists (Plag and DiGiacomo, 2014). The GCZIS will enable routine, scientifically sound assessments of the condition of coastal ecosystems across the land-sea interface on local to global scales, covering both developed and developing regions. It will also provide the information needed to understand state and trends in coastal communities and their social, economic and environmental conditions. It will serve as a resource for all other components of the Blue Planet task. The initial GCZIS development will be focused on study cases for Virginia (United States), the Caribbean, and West Africa. Other regions, where there is strong user interest and capacity for implementation, can also participate as study cases. Issues to be addressed include:

- Identification of key indicators of ecosystem and socio-economic system condition;
- Data requirements for computing these indicators;
- Observing system requirements for provision of indicators of integrated coastal system assessments as needed by decision makers;
- Suitable web-based solutions to convey and deliver information in a timely manner; and
- Mechanisms enabling contributions from "citizen scientists" and through "crowd sourcing," which are of particular relevance in regions with gaps in monitoring infrastructure.

Toward developing monitoring and forecasting services of the type described above, a Sea Level Rise Community of Practice (SLRCoP) has recently been established, joining and complementing other existing CoPs. The SLRCoP is open to all stakeholders concerned with the challenge that changing sea levels pose to large parts of the coastal zone, particularly in urban areas. Based on the notion that risk can be described as the product of hazards probability, vulnerability of infrastructure and social fabric, and exposure, the SLRCOP aims to quantify the probability of sea-level related hazards, understand the vulnerability of coastal infrastructure, ecosystems, and communities with respect to these hazards, and assess the risk based on the exposure of social, environmental, and economic assets to hazards. The resulting sea level risk maps provide a basis for coastal land use planning, adaptation, and mitigation. More information can be found at: http://www.slrcop.org/. A related effort is under development for the monitoring of coastal and inland water quality, particularly via remote sensing.

International Cooperation

The CZCP and its contributors and stakeholders have active liaisons with many organisations, including: GOOS, GTOS, GCOS, the Global Geodetic Observing System (GGOS) and DIVERSITAS (an international programme of biodiversity science). Nested within these are links to a broad diversity of global and regional coastal and ocean observing institutional and programmatic networks, e.g., GRAs, POGO, World Association of Marine Stations, IOC-IODE Ocean Data and Information Networks, ChloroGIN, GODAE OceanView, GEO Biodiversity Observation Network (GEOBON), Global Alliance of CPR Surveys, Global Terrestrial Network for River Discharge, and the Global Coral Reef Monitoring Network. International cooperation and coordination across these and other coastal and ocean stakeholders, in conjunction with GEO Participating Organisations and related entities, will be essential to implementing and realising the benefits of the GEO Blue Planet activities (particularly those efforts led or supported by the CZCP). For example, the GCZIS will be developed in cooperation and consultation with relevant international communities that have already developed and implemented related systems, e.g., the International Coastal Atlas Network (ICAN).

Links to Societal Benefits

GEO has been constructing GEOSS on the basis of a 10-Year Implementation Plan for the period 2005 to 2015. The Plan defines a vision statement for GEOSS, its purpose and scope, expected benefits, and the nine Societal Benefit Areas (SBAs), Disasters, Health, Energy, Climate, Water, Weather, Ecosystems, Agriculture and Biodiversity. Oceans and coastal zones cross-cut and intersect, directly or indirectly, with all nine SBAs. This highlights their particular importance and critical value from a socio-economic perspective. In this context, the CZCP focuses both on research, practical applications, and practice-relevant knowledge related to coastal zone management and ecosystem-based approaches for sustainable development. It interacts directly with a number of GEO Participating Organisations and the SBIB.

GEOSS will yield a broad range of societal benefits, notably: reducing loss of life and property from natural and human-induced disasters (e.g., assessing and forecasting vulnerability to coastal flooding); understanding environmental factors affecting human health and well-being (e.g., monitoring and predicting exposure to waterborne pathogens); improving the management and availability of energy resources (e.g., siting and maintenance of coastal wind farms); understanding, assessing, predicting, mitigating, and adapting to climate variability and change (e.g., understanding the ecosystem impacts of ocean acidification); improving water resource management through better understanding of the water cycle and impact of terrestrial loadings (e.g., coastal eutrophication and hypoxia); improving weather information, forecasting and warnings (e.g., improved hurricane forecasts); improving the management and protection of terrestrial, coastal and marine ecosystems (e.g., HAB forecasts, improved fisheries management and enabling food security); supporting sustainable agriculture and combating desertification (e.g., mitigating habitat loss and modification); and, understanding, monitoring and conserving biodiversity.

As already identified, high priorities for the CZCP related to GEOSS currently include, but are not limited to, improved forecasts of sea-level rise and the associated increase in frequency and extent of coastal inundation, as well as monitoring changes in water quality and associated impacts on ecosystem health and productivity.

Conclusion

Sustainable development of the coastal zone represents one of the great challenges presently facing humanity given the tremendous socio-economic and ecological importance, as well as complexity, of these regions. In fact, the sustainability of the coastal zone is one, if not the, frontline in humanity's sustainability crisis (Glavovic, 2013). Conducting workshops and generating community reports is relatively "easy"; the primary challenge is providing effective, sustained support for users and implementation and continuity of observing capabilities enabling deliberative governance and integrated, ecosystem-based approaches for coastal management and decision-making. CoPs typically develop on a best-effort basis with extremely limited (human and other) resources. Adequately supporting the wide range of users and services that could benefit from improved access to and utilisation of Earth observations remains a significant challenge. CZCP looks forward to broadening its group of participants, stakeholders and partners, which can facilitate increasing the support provided to users and helping to advance coastal applications and services across the globe. In the coming years, the CZCP intends to:

- Support implementation of the GEO Blue Planet task, particularly the *Services for the Coastal Zone* component;
- Support implementation of coastal GOOS, in particular facilitating recommendations made by the GOOS Panel for Integrated Coastal Observations (PICO-I, 2012);
- Gain broader geographic representation and user participation, and facilitate enhanced communication, training and outreach activities;

- Support and engage users through timely coastal zone assessments, and by facilitating diverse applications and services;
- Develop, implement and populate the GCZIS;
- Facilitate the development of user-driven pilot projects though community workshops, capacity building activities and support broader stakeholder linkages and coordination efforts.

CHAPTER EIGHTEEN

COASTAL OCEAN COLOUR OF AUSTRALIAN WATERS: PROGRESS AND OUTLOOK

ANDREW D.L. STEVEN, VITTORIO E. BRANDO,
LESLEY CLEMENTSON, ARNOLD G. DEKKER,
NICK HARDMAN-MOUNTFORD,
JONATHAN HODGE, EMLYN JONES,
EDWARD KING AND THOMAS SCHROEDER

Introduction

With a continental coastline of ~ 36000 km that spans 35° of latitude, Australia's coastal waters vary greatly from clear ocean to turbid, optically-complex waters (Figure 18-1). With the third largest exclusive economic zone in the world, Australia's environmental responsibilities extend beyond its coastal waters to its oceans and territories, including almost half of Antarctica. Most of this area is difficult to access, and information can only be cost-effectively gathered by using satellites. Geosciences Australia (GA), the Bureau of Meteorology (BoM), and the Commonwealth Scientific and Industrial Research Organisation (CSIRO) are the Australian Government agencies jointly responsible for Australia's civilian EO programme. These agencies, together with a broader Ocean Colour (OC) community, have worked cooperatively to develop OC-based products and services for a range of end-users. Engagement has mostly been with environmental management agencies, mainly for environmental monitoring and enforcement.

The approach taken in the development of reliable and useful OC products is one of regional validation and, as required, algorithm development, to produce user-defined products supported and delivered by a national operational infrastructure. Since 2009, the development and application of OC has been significantly advanced through major national investments, including the National Collaborative Research Infrastructure Strategy (NCRIS) that encompasses the Integrated Marine Observing System (IMOS), the Terrestrial Ecosystem Research Network (TERN), and the Australian National Data Service (ANDS) (http://ncris. innovation.gov.au). These initiatives have supported investments in infrastructure for *in situ* validation, atmospheric correction, and computing, as well fostering EO communities-of-practice.

Regional validation has been most significantly advanced in the waters of the Great Barrier Reef (GBR), and some quasi-operational OC products are now delivered to the responsible management agency, the GBR Marine Park Authority, via the BoM. Under a significant 5-year multi-agency collaboration called *eReefs* (http://www.ereefs.org.au/), further regional validation of GBR waters, operational delivery, and assimilation with hydrodynamic and biogeochemical models are being pursued.

Nationally, while there is interest in the potential of OC to provide environmental services, adoption in other regions of Australia has been relatively slow. There is, however, a renewed momentum in the application of EO to Australia's environmental needs, a key priority being to develop a National Earth Observations from Space Infrastructure Plan (Australian Government, 2013), which scopes the role Australia should be playing regionally and internationally.

In situ observing data

The IMOS Satellite Remote Sensing facility supports OC calibration and validation (Cal/Val) through collection of ground observations at the Lucinda Jetty Coastal Observatory (LJCO) and from Australia's Marine National Facility, the Research Vessel (RV) Southern Surveyor.

LJCO was established in 2009 at the end of the 5.8 km long Lucinda Jetty in the coastal waters of the central GBR, to acquire two complementary high-frequency data streams. An autonomous above-water radiometer (CIMEL SeaPRISM) measures the water-leaving radiance and acquires atmospheric measurements for retrieving aerosol optical properties. An *in situ* underwater instrument package is deployed to characterise the inherent optical properties (IOPs) of these complex coastal waters. Details on all instruments and sampling protocols are provided in Brando *et al.* (2010).

The deployment of ship-based above-water radiometry aims to evaluate atmospheric correction methods over a wide range of water types.

For this purpose, the "Dynamic above water radiance and irradiance collector" (DALEC) has been deployed since July 2011 on the RV Southern Surveyor to collect extensive transects of above-water reflectance. The data collected to date (see tracks in Figure 18-1) is available through the Australian Oceanographic Data Network (AODN) portal (http://data.aodn.org.au/IMOS/opendap/SRS/SRS-OC/SRS-OC-SOOP _Rad/).

IMOS also supports the collection of further bio-optical data by coupling *in situ* sensor systems and monthly water sampling in a continental scale network of coastal monitoring stations, the National Reference Stations (Lynch *et al.*, 2013), as well as by a fleet of autonomous underwater ocean gliders that undertake measurements from within shelf and boundary currents in Australian waters. To ensure consistency across the bio-optical data streams collected on stationary and mobile platforms, a common calibration procedure has been defined (Earp *et al.*, 2011).

The optical data collected from these platforms, together with discrete *in situ* observations collected since 1997, are collated into a bio-optical database containing more than 11,000 bio-optical observations (Figure 18-1), available through the AODN (http://data.aodn.org.au/IMOS/opendap/ SRS/BioOptical/). All observations are transferred to international space agencies, such as NASA and ESA, for inclusion in their global databases, SeaBASS and MERMAID, respectively. The availability of such data to international space agencies will ensure that the validity and accuracy of global algorithms developed for future sensors should improve for Australian waters.

Algorithm development and validation in the Great Barrier Reef

Regional OC algorithm development has focussed on coastal waters of the GBR to underpin quasi-operational environmental reporting. In the GBR, spatial and temporal variability in the composition and concentration of particulate and dissolved matter results in significant variability in the IOPs (Blondeau-Patissier *et al.*, 2009; Oubelkheir *et al.*, 2006). In these waters, the accuracy of chlorophyll estimates from seven SeaDAS-implemented global chlorophyll algorithms generally degrades rapidly with increasing concentration of particulate and dissolved matter (Qin *et al.*, 2007). In addition, frequent standard atmospheric correction (AC) failure is observed above highly turbid or absorbing coastal waters, resulting in negative water-leaving reflectance.

To improve the accuracy of water quality and IOP estimates in GBR coastal waters, two physics-based inversion algorithms were developed and coupled: one addressing the AC and air–water interface, and the other the in-water constituent retrieval (Brando *et al.*, 2012, Schroeder *et al.*, 2012). The AC algorithm applies radiative transfer simulations and Artificial Neural Network (ANN) inversion to derive remote sensing reflectance at mean sea level. For water masses mainly influenced by detrital and mineral material, the ANN algorithm has been shown to perform more accurately compared to other SeaDAS-implemented AC methods (Goyens *et al.*, 2013) and, as such, improves water quality inversion over turbid flood plumes in the GBR (Figure 18-1).

The IOPs and concentrations of optically active constituents are then estimated using an adaptive Linear Matrix Inversion, aLMI, (Brando *et al.*, 2012) developed to deal with substantial spatial and temporal variability in optical properties, particularly between the tropical wet and dry seasons due to the influx of river water during flood events. A comparison of MODIS Aqua retrievals of chlorophyll, CDOM and total suspended solids with concurrent *in situ* observations showed that the aLMI in-water constituent retrieval algorithm coupled with the ANN AC led to lower uncertainties than all other SeaDAS implemented water quality inversion algorithms in the GBR (Figure 18-1 and Brando *et al.*, 2013). This significant improvement has provided regional-scale confidence in coastal OC products for uptake in environmental monitoring and reporting, and has led to them being used as assimilation inputs to constrain hydrodynamic, sediment transport and biogeochemical models in the GBR region.

Integrating observations and models: Data assimilation

Satellite-derived measurements of the surface layer of the ocean and *in situ* subsurface data obtained from various platforms are being used to better constrain biogeochemical (BGC) models representing the cycling of key nutrients through the lower trophic levels of Australian marine ecosystems. As these BGC models have often relied on simplistic parameterisations of key eco-physiological processes, a key challenge is to merge observations of the 4-dimensional (4D) evolution of the water column, with simplified representations of reality encapsulated in BGC models (Parslow *et al.*, 2013). The quantitative use of observational data to constrain the model state and parameters is often referred to as Data Assimilation (DA), and its application to marine BGC models is discussed in Matear *et al.* (2011).

Due to the different sources of error in marine BGC models (compared with the hydrodynamics), alternative approaches to DA, including Bayesian Hierarchical Modelling (BHM) (Parslow *et al.*, 2013) and emulator approaches (Margvelashvili *et al.*, 2013; Mattern *et al.*, 2012), have been successfully applied to marine sediment and BGC problems. Both of these methods can be used for joint inference on the model state and parameters. When these algorithms are combined with recent advances in the formulation of coastal BGC models, i.e., calculating both the Inherent and Apparent Optical properties within the BGC model (Baird *et al.*, 2013), a more diverse range of OC products can be assimilated. BGC DA systems can now combine models and observations to provide quantitative estimates of model uncertainty which is required by policy makers and mangers for risk-based decision-making.

Operational delivery and products

The R&D stimulated over the past decade by readily available, high quality data from the MODIS and MERIS sensors also led to the development of operational processing and analysis techniques for coastal water condition monitoring from space. With MERIS ended, and MODIS near the end of its mission, substantial efforts will be directed to transition to the recently launched United States VIIRS sensor and the forthcoming OLCI instrument on the ESA Sentinel missions to ensure long-term data continuity. This will require substantial retuning and recalibration of the existing processing and analysis system and programmes that can blend data from both instruments are most likely to capitalise on the recent progress.

Great strides have been made towards adopting a national approach to manage the satellite data archives. Historically, the diversity of coastal environments has resulted in the growth of multiple, often incompatible satellite data archives covering different areas, obstructing data exchange and algorithm comparison. Recent national *eResearch* infrastructure investment has delivered compute-rich facilities, co-located with large data stores, to the research community. This opportunity has been exploited, through national partnerships such as IMOS and TERN, to construct comprehensive satellite data archives with national coverage. This reduction in duplication also provides a consistent data platform for the whole continent, making possible the progressive extension of existing regional monitoring systems with a primary focus on robust science-based algorithms.

Well-organised, national-scale satellite data archives play another key role, providing a foundation for the deployment of data standards that underpin interoperability across the coastal-information system field. Rapid advances in geospatial metadata tools, display and analysis systems and, most recently, provenance-aware workflows, can be promulgated very effectively through adoption in key data sets. In the *eReefs* partnership, core data and metadata standards and practices are utilised from the base satellite data level to build a flexible multi-tiered environmental information system. As more data sources (primarily *in situ*) come online, and with two-way exchange of information with hydrodynamic models, this system exemplifies the benefits of modern data interoperability practice, delivering a range of near real-time and historical observations (e.g., chlorophyll, sediments and Secchi depths) and derived information products, such as water quality compliance, both at native sampling resolution and aggregated to a variety of regions and timescales relevant to managers.

Links to Societal Benefits

In Australia there is a growing operational and economic dependence on EO data for a diverse range of applications including modelling climate, forecasting weather, monitoring water management and quality, surveillance of oceans, mapping forests, estimating agricultural production, mitigating hazards, responding to disasters, assessing urban expansion, locating mining and energy resources, maintaining national security, protecting borders, positioning, transport and navigation (Geoscience Australia, 2011, CSIRO 2012). This currently involves at least 92 major programmes (Geoscience Australia, 2010) and $3.3 billion per year GDP contribution for both direct and indirect productivity measurements (ACIL Tasman, 2010).

While there is general interest in the potential of OC to provide environmental services, adoption elsewhere (e.g., fisheries) has been relatively slow. For example, in the GBR region the *eReefs* project brings together research, operational, and management agencies, allowing iteration and feedback between end-users and developers of products that, in turn, leads to refinement of the ocean colour products available, including water quality guideline compliance maps and a dashboard of conditions, flood plume products and algal bloom assessment. The value from investments in EO of river plume information alone has been estimated to generate up to 37 million AUD/year in benefits (Bouma *et al.,* 2011).

Supporting operational OC services requires on-going dialogue between the scientific disciplines involved in developing new approaches and products and the stakeholders finding new areas of application for these products, whether in applied research, environmental monitoring or operational prediction. Within Australia, bio-optical and remote sensing scientists engage with each other through the IMOS Bio-optical working group. From late 2012, this group opened its activities to a wider group of scientists covering the spectrum of OC research, and an OC community-of-practice is emerging.

The federal government's development of a National Earth Observation Infrastructure Plan and CSIRO's establishment of the Earth Observation and Informatics Transformational Capability Platform provide further opportunities for the OC community to interact and exchange knowledge with other communities working with different EO sensor-types and across Earth system domains (atmosphere, terrestrial vegetation, and minerals). These communities provide cross-fertilisation of ideas and approaches, helping ensure strong adoption of cutting-edge developments.

International Cooperation

Australia recognises that stronger international relationships improve the relevance of satellite information to Australia. It is also recognised that these Australian activities exist within a wider international context, not only from the perspective of access to EO data, but also the imperative to share knowledge and collaborate at a global scale. Such collaboration is manifested, for example, through active participation in IOCCG and GEO working groups (e.g., inter-comparison of Retrieval Algorithms for Coastal Waters, GEO Working Group on Inland and Near-Coastal Waters), space agency validation teams (e.g., for Sentinel-3) and the NASA sponsored SeaWiFS HPLC Analysis Round-Robin Experiments (SeaHARRE).

CSIRO has been nominated to be the Chair of CEOS in 2016 and is now considering a range of initiatives that could be implemented to foster closer regional cooperation and operational capability, including hosting of the GEO Blue Planet Symposia in 2014. There is also clear recognition of the need for greater collaboration to better understand the dynamics of the Indian Ocean region.

Conclusion

Space-based Earth Observation is the only way to effectively monitor Australia's environment and is an important area of innovation for Australia. Over the last five years, significant investments in bio-optical and compute infrastructure, greater coordination of the EO research community and end-user engagement, and on-going support from responsible EO institutions, coastal management agencies, and the broader research community have established solid foundations for Australia's stronger future utilisation of EO.

Looking forward from 2013, key drivers will be the development and implementation of the National Earth Observations from Space Infrastructure Plan (Australian Government, 2013), a stronger focus on the development of operational oceanography capability around Australia, and the need for better surveillance and forecasting capability of natural and man-made disasters as well as maritime operations.

A key task for the Australian EO Community will be the southern hemisphere validation of the many new satellite sensors that are to be launched over the next few years, and the development of customised OC products (e.g., primary productivity) for an increasingly sophisticated and diverse range of end-users who rely on these products to guide decision making.

Efforts to build a next-generation suite of marine hydrodynamic, BGC, and ecosystem models with better forecasting capability for Australia's coasts and oceans will require better access to, and assimilation of, these satellite data-streams. Close collaboration between the institutions responsible for developing and implementing EO in Australia will make this a reality.

Acknowledgements

Many people have made significant and on-going contributions to the development of ocean colour capabilities in Australia and we especially want to acknowledge the leadership of Stuart Phinn, Alex Held, Merv Lynch and Tim Moltmann. Many of the activities described here have been possible due to resources made available through CSIRO Wealth from Oceans Flagship and the Earth Observation and Informatics Transformational Capability Platform; the Australian Federal Government Reef Rescue Marine Monitoring Program; National Collaborative Research Infrastructure Strategy Funding and *eReefs* partner organisations.

CHAPTER NINETEEN

SIMCOSTA:
BRAZILIAN COASTAL MONITORING SYSTEM

MILTON KAMPEL

Introduction

Ocean time series provide vital information needed for assessing ecosystem change (Church *et al.*, 2013). Regular and recurrent observations of a suite of core parameters at fixed locations are important as they enable the analysis and interpretation of physical, biogeochemical and ecological variability within these locations. This knowledge contributes to our understanding of the effects of global climate change, characterising temporal scales of variability from months to decades. Long-term time-series measurements are crucial for isolating secular trends in the ocean carbon cycle from natural variability, and for determining the physical and biological mechanisms controlling the system. Year-to-year variations in physics (e.g., upwelling, downwelling), bulk biological production and ecological shifts (e.g., community structure) can drive significant changes in ocean carbon cycling and ecology. The biological and chemical responses to natural (e.g., El-Niño Southern Oscillation – ENSO, dust deposition events) and anthropogenic perturbations are particularly important with regard to evaluating the prognostic models used in future climate projections.

The science community, policymakers, and society need an observing system for the global climate and ecosystem in order to detect changes, to describe and quantify them, to understand and explain them and to develop a capability to predict them. Oceanic and coastal time series are therefore an invaluable tool for marine scientists and have been established in many locations around the world since the beginning of the 20th century (the majority having been initiated from the 1950s onwards). These time series were initially concentrated in coastal regions as they were operated by marine institutes located along the coastline. These time series generate a range of data usually including temperature and salinity as a minimum; many also collect biological samples to study the abundance and diversity of marine organisms, such as phyto- and zooplankton, fish, or benthic organisms. In combination, these data facilitate studies not only related to how these assemblages are affected by their physical and chemical environment, but also on how the environment shapes biological interactions over the long term.

On-going anthropogenic climate change has certainly increased the scientific interest in time-series analyses and their importance is increasingly recognised at the political level. Data centres all over the globe are now concerned with the long-term preservation of newly generated data and the retrieval of historic data sets, a task that has not yet been completed. The challenge for scientists around the world now is the integration and analysis of these often complex data sets. POGO has compiled a list of time-series stations (see Chapter 22 and http://ocean-partners.org/ocean-observations/long-term-datasets), many of which are members of OceanSITES (see Chapter 5 and http://www.oceansites.org).

In-situ Coastal Monitoring

Remote sensing information provides an invaluable tool to obtain synoptic quasi-continuous information on the surface ocean. Nevertheless, in order for these observations to be analysed and processed into reliable information for managers and decision makers, the retrieved signals must be validated to ensure they represent actual properties of the ocean within certain confidence limits. This is even more critical when satellite products are used in models to provide ecological indicators, whose behaviour are known to change with environmental conditions; and hence should be tested regionally. The only way to attain information for these validation exercises is through the development of *in situ* time-series studies in different regions. Furthermore, *in situ* work allows sampling of the entire water column, as well as more elaborate studies (e.g., biodiversity, plankton dynamics, physiology) which are not possible through remote sensing alone (Sathyendranath, 2010).

Several *in situ* time-series stations form part of Antares (see Chapter 11 and http://www.antares.ws), the Latin-American node of ChloroGIN, and the initial network around which ChloroGIN has been built (see Chapter 12 and http://www.chlorogin.org). Within this context, Brazil has established a national buoy programme – PNBOIA, to implement an array of drifters and coastal moorings, both monitored by satellite, whose data are made available to the scientific community in near-real time.

International Cooperation

The Brazilian Coastal Monitoring System (SiMCosta) project is a collaboration between five Brazilian research institutes (Instituto Nacional de Pesquisas Espaciais – INPE; Universidade Federal do Paranà – UFPR; Universidade Federal de Santa Catarina – UFSC; and Centro de Biologia Marinha Universidade de São Paulo – USP), one Canadian university (Dalhousie University) and a number of Brazilian and Canadian private companies. The project, funded by the Brazilian Federal Ministry of Environment (MMA) and the Climate Change National Fund (Fundo Nacional sobre Mudança do Clima), includes a number of initiatives designed to plan, acquire, and develop the equipment and knowledge capacity needed to implement a robust monitoring programme off the Brazilian coast.

SiMCosta aims to improve the Brazilian coastal observing system by using advanced and integrated methodologies to obtain, analyse, and distribute permanent and continuous temporal series of climatic and oceanographic parameters collected at several points along the coast. SiMCosta seeks to detect long-term trends, improve the capability of predicting the effects of climatic variability and climate change, and contribute with other nations to early-warning systems for extreme events. In the medium term, SiMCosta aims to install platforms along the entire Brazilian coastal region. At this early stage, it will focus upon the Southern Brazilian Region, covering approximately 3,000 km of the coast to generate a reliable data base on the Oceans and Coastal Zone for modelling, forecast and alert purposes.

The SiMCosta project was initiated in December 2011, in response to a specific request from the Ministry of Environment (MMA) and a long-standing need identified by the Brazilian scientific community. It was initially financed with resources from the Climate Change National Funds (MMA/ Fundo Clima), to a sum of R$ 1.96 million. The Coastal Zones Network, linked to the Brazilian Network for Global Climate Change Studies (Rede CLIMA) and the National Institute of Science and Technology for Climate Changes (INCT for Climate Changes), will be responsible for establishing and implementing the monitoring system. The Federal University of Rio Grande (FURG) and its Institute of Oceanography (IO) are national leaders in oceanic and coastal sciences. The IO is the headquarter of several integrated Brazilian programmes and initiatives, including the Coastal Zone Network, the National Institute of Science and Technology for the Sea (INCT-Sea), and the Brazilian Long Term Ecological Research (Brazilian LTER). SiMCosta is now proposing and defining the oceanographic sensors and instruments, which must be coupled with the meteorological buoys.

Specific objectives of the Brazilian Coastal Monitoring System include:

- To map vulnerabilities of the coastal zone;
- To anticipate physical, biotic, and socioeconomic impacts;
- To generate future scenarios;
- To evaluate mitigation alternatives;
- To alert occurrence of extreme events;
- To preview processes linked to climatic effects; and
- To identify trends over time.

In addition to SiMCosta and PNBOIA, there are other ongoing initiatives along the Brazilian coast with complementary objectives and similar approaches, including the Ocean network also under the National Institute of Science and Technology for Climate Changes (see http://inct. ccst.inpe.br/index-in.php), and the National Institute of Science and Technology for the Sea – Centre for Integrated Oceanography (see http://www.inctmar.furg.br/). An integrated vision is foreseen in terms of an initial phase and a medium term (see Figure 19-1).

The configuration of the SiMCosta buoys includes a meteorological system measuring: Pluviometric precipitation, atmospheric pressure, wind direction, wind speed, relative humidity, solar radiation, air temperature, and CO_2 concentration. Oceanographic data to be collected at the surface includes backscattering, chlorophyll, coloured dissolved organic matter, nitrate, dissolved O_2, depth, pH, salinity, SST, and turbidity. Subsurface oceanographic data include waves and current profiles. The flotation system is based on the Tropical Atmosphere (TAO) TRITON buoy, and the system also comprises a radio communication system for oceanographic data telemetry, as well as a satellite system for meteorological data transmission. A dedicated software package was developed for end-user data access on the Internet.

Links to Societal Benefits

By observing the ocean and coastal regions of Brazil, SiMCosta will provide further in-depth knowledge on how these ecosystems function, and how they might respond to climate change. Changes in ocean temperature impact on many aspects of the oceans, including major current systems, the intensity and the frequency of storms. These will have a profound effect on the large population residing along the coasts that is dependent on the ocean for its income and survival. Clearly, ocean observations such as the ones acquired through the SiMCosta system have the potential to produce societal benefits in all nine categories defined by GEO (see Chapter 2).

Conclusion

Historically, oceanographers made great discoveries by conventional spatial exploration – they travelled to new places in the oceans and discovered unexpected phenomena that advanced our understanding of a particular process (Clark and Isern, 2003). Today, important discoveries are often the result of long time-series data sets.

Time is so fundamental to our understanding of Earth system processes that it is sometimes ignored in the development of conceptual models of how ocean habitats are structured and how they function (Karl, 2010). Oceanic ecosystems are complex, time variable, non-steady state, with nonlinear features, and need to be properly approached. However, under-sampling or scarcity of data is a reality in many parts of the world ocean (Platt *et al.*, 1989) and still constrains the interpretation of available field data.

The Coastal zones are vulnerable to the impacts of global climate change, immediately exposed to rising sea level, the elevation of air temperature and sea surface temperature, ocean acidification, changes in wind regimes and discharges of rivers, and to extreme events (Trenberth *et al.*, 2007, Bindoff *et al.*, 2007). In Brazil, research related to impacts, vulnerability, and adaptation to climate change are limited by the shortcomings of knowledge about the natural dynamics of coastal ecosystems and the lack of long time series (Copertino *et al.*, 2010). In this context, it becomes essential to create and maintain permanent multidisciplinary and efficient observational systems, enabling analyses on several scales of time and appropriate space.

SiMCosta therefore aims to strengthen oceanic observation systems in strategically selected regions of Brazil with advanced, robust, and integrated methodologies to generate continuous time series of climatic, hydrological, and oceanographic variables.

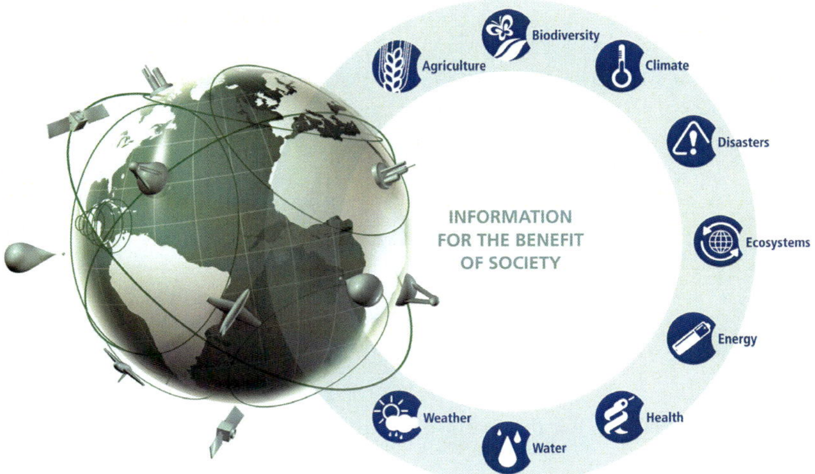

Figure 2-1: The nine Societal Benefit Areas of GEO

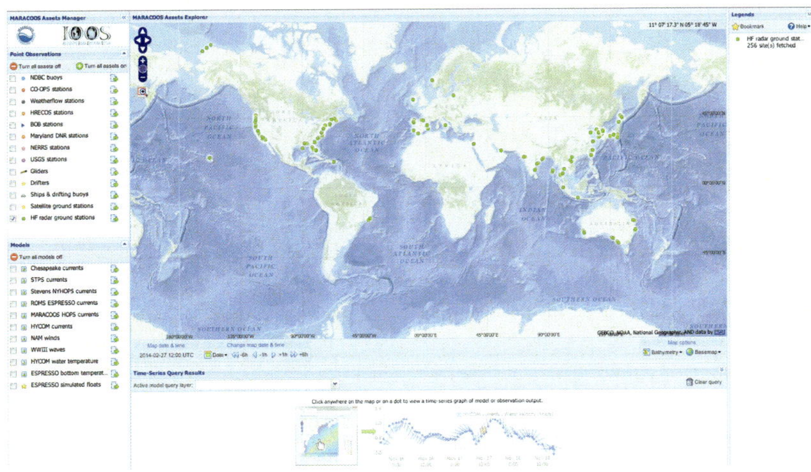

Figure 4-1: Representative sample of global HF radar

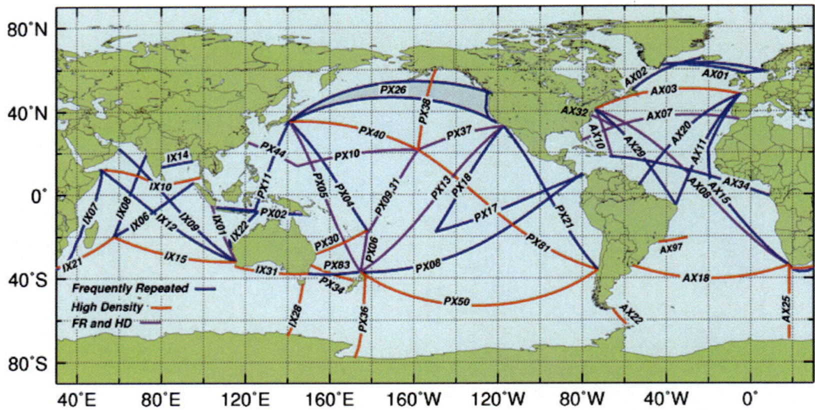

Figure 7-1: The current XBT network containing OceanObs99 and OceanObs09 recommendations

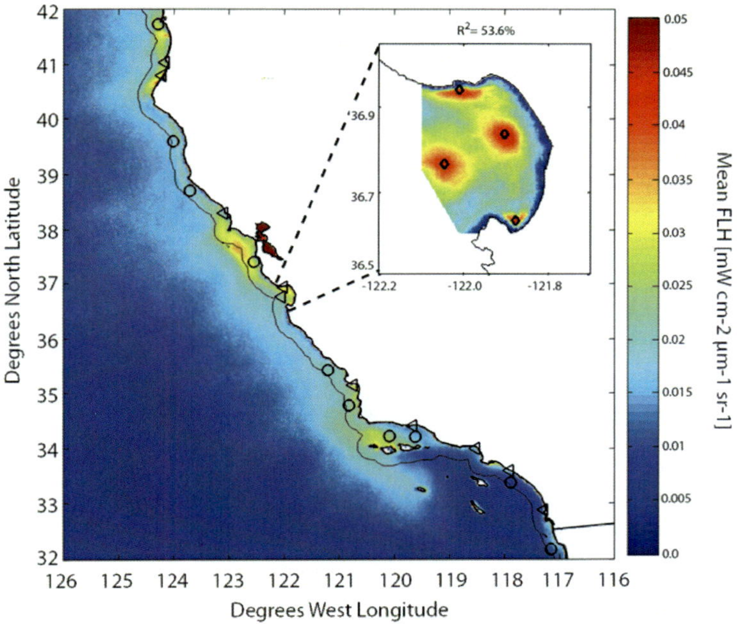

Figure 8-1: Proposed (circles) and existing (triangles) HAB monitoring stations covering the California, USA, continental shelf, overlaid on climatological fluorescence line height (FLH). These optimal sites, in combination with satellite remote sensing, would capture 80+% of HAB events. Inset: variance of FLH (a proxy for high biomass HAB events) explained by existing moored and shore-based HAB stations for Monterey Bay, CA. While ~54% of the variance is captured by four stations, there is a distinct spatial bias, alleviated by combining observational sites with remote sensing and modelling. Figure adapted from Frolov *et al.* 2013

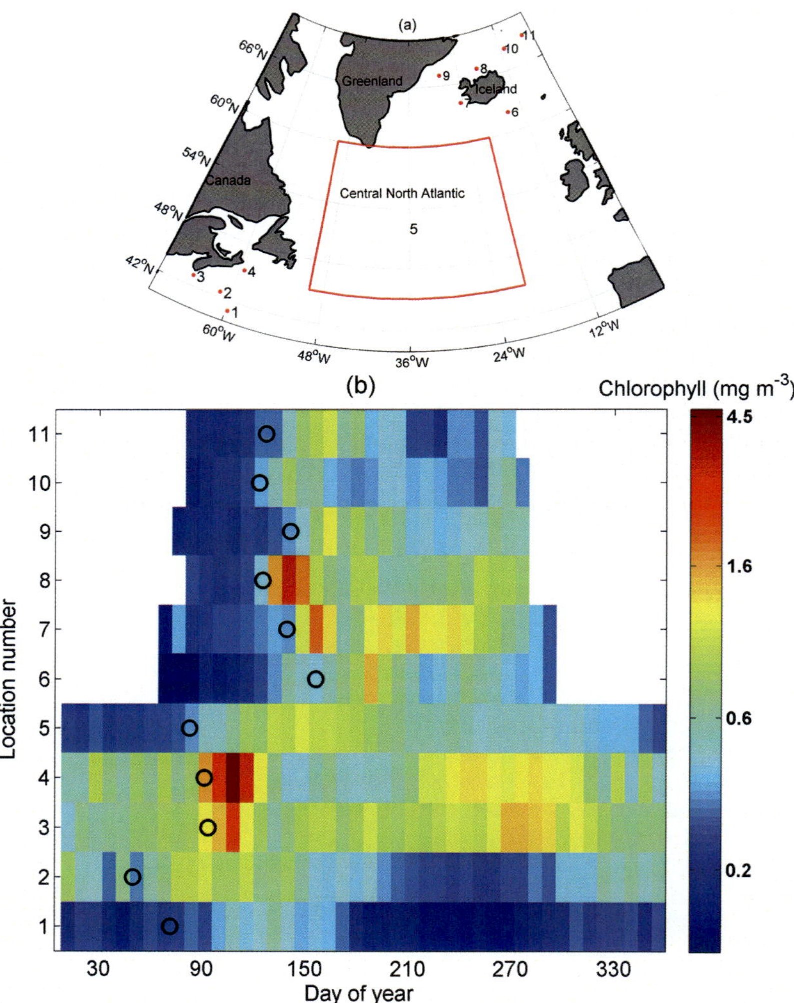

Figure 10-1: (a) The study region with 11 geographic locations in the North Atlantic. 1 Gulf Stream, 2 Slope water, 3 western Scotian Shelf, 4 eastern Scotian Shelf, 5 Central North Atlantic, 6 northern North Atlantic, 7 southern Iceland Shelf, 8 northern Iceland Shelf, 9 Polar water, 10 Arctic water, 11 mixed water which is the recirculated Atlantic water and is much influenced by the Arctic water masses. (b) The seasonal and geographic variation of climatological 8-d chlorophyll concentration. Black circles indicate the time of spring bloom initiation

Figure 11-1: Location of the time-series field stations of the Latin-American Antares network

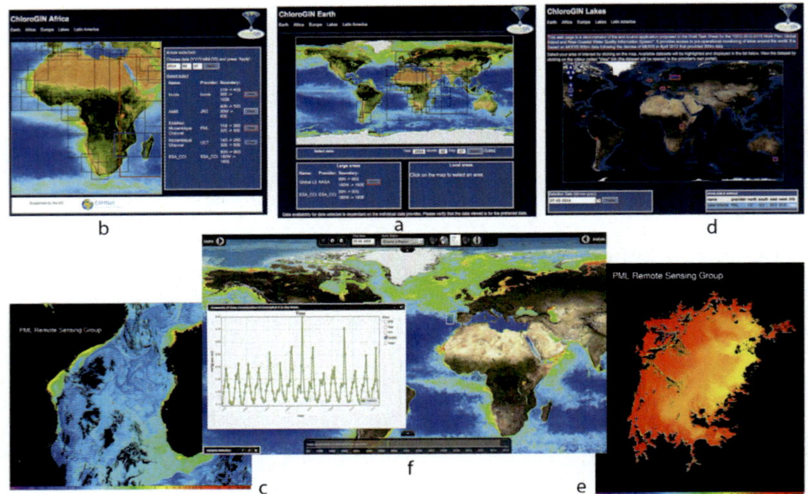

Figure 12-1: a) ChloroGIN global coverage ; b) ChloroGIN Africa portal showing data availability for Mozambique channel; c) Example NASA Aqua-MODIS chlorophyll-a image for Mozambique channel, 27 July 2013; d) ChloroGIN lakes portal; e) example ESA Envisat MERIS chlorophyll-a image for Lake Victoria, 16 Feb 2012; f) OpEc visualisation and analysis portal showing time series for 1997-2012 of chlorophyll-a concentration to the west of Portugal extracted from OC CCI version 1 global data.

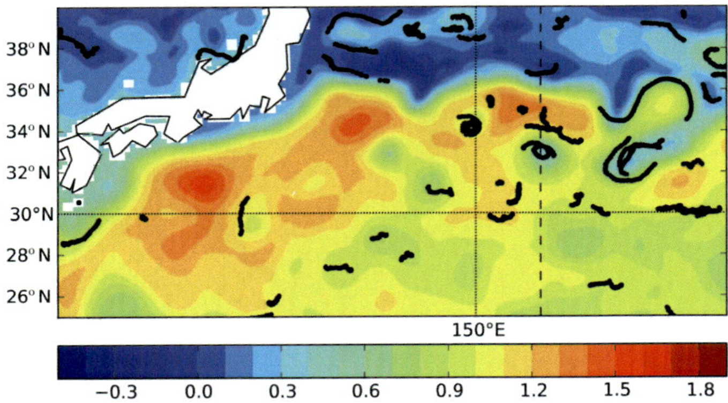

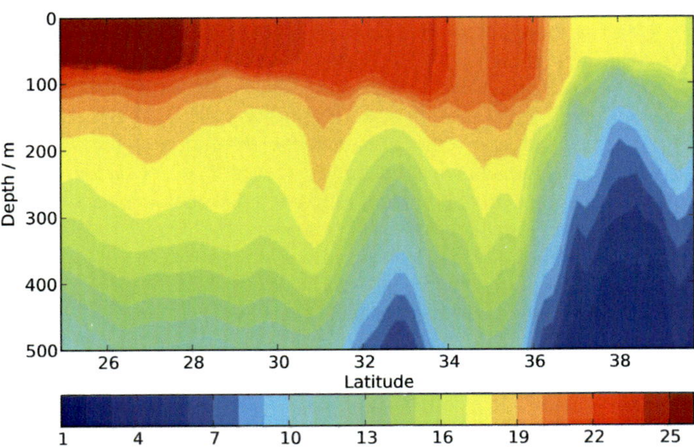

Figure 13-1: (top): The sea-surface height field (SSH in m) for the south coast of Africa from the global FOAM system on 21 Nov 2011 with the locations of surface drifters from 18-24 Nov 2011 overlaid. (Bottom): The temperature as a function of depth on the same day along the section through 40°E marked (dashed line) on the top plot. Graphics courtesy of Jennie Waters (Met Office)

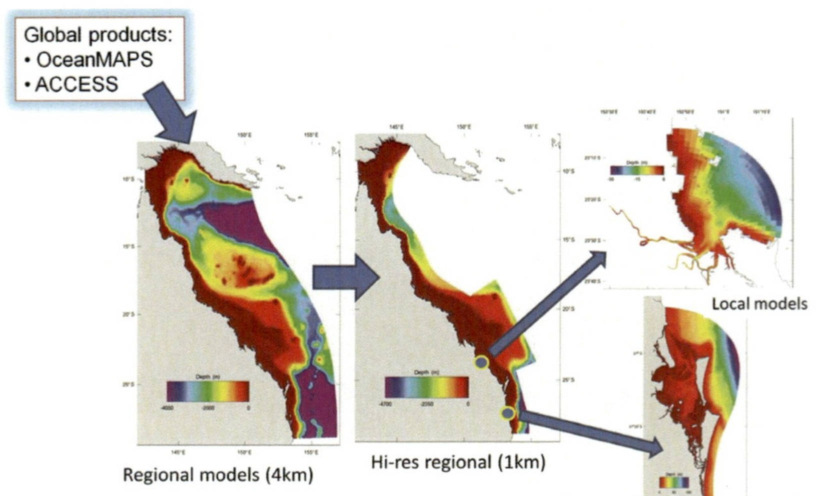

Figure 14-1: A nested approach to traverse scales (Herzfeld, CSIRO, pers. comm., 2013). The colours show bathymetry. The regional configurations span a portion of the Coral Sea between Papua New Guinea and Queensland, Australia. The local models cover the Fitzroy River estuary and Moreton Bay off Brisbane. The figure shows the downscaling cascade required to achieve resolution at the local scale while maintaining open boundary nesting ratios of less than 5:1. Often multiple nests are required to achieve these boundary ratios – in this case 2 nests at 4km and 1km scales. The downscaling starts with global models, e.g., OceanMAPS (Oke *et al.*, 2008) at 10 km resolution for the ocean initial and boundary conditions, and ACCESS http://www.bom.gov.au/nwp/doc/access/NWPData.shtml for surface fluxes. Local models can then conceivably have resolution of ~200m at their open boundaries. Figure provided by Mike Herzfeld, CSIRO CMAR and the eReefs project

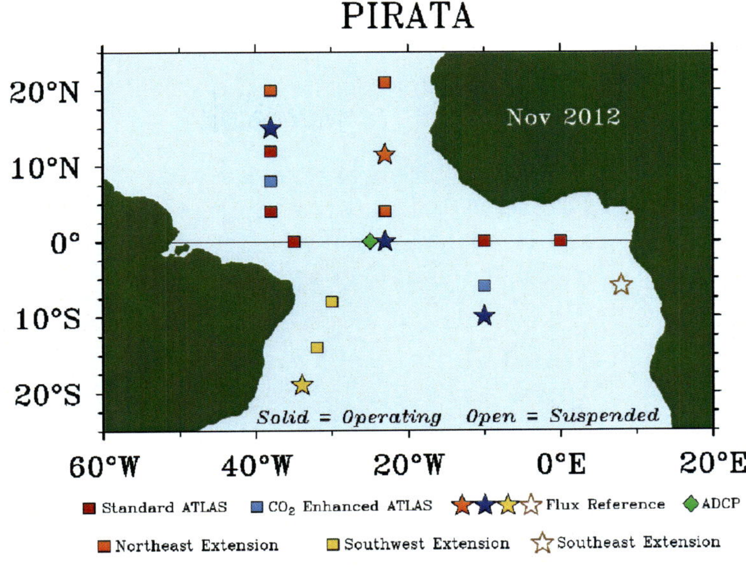

Figure 15-1: The PIRATA array (from http://www.pmel.noaa.gov/pirata)

Figure 17-1: In many areas, the coastal zone is challenged with a range of conflicting uses, including residential and industrial built environment in urban settings, transportation on land and water and in the air, harbours, fishing and aquaculture, agriculture, recreation and tourism, and preservation of coastal ecosystems. The San Francisco Bay area combines all of these uses in a dynamic setting (Photo: Hans-Peter Plag, 24 September 2010)

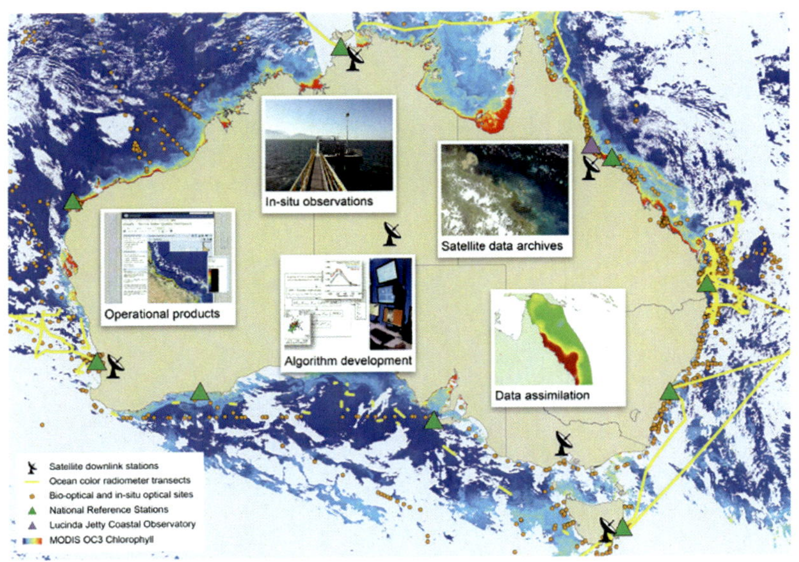

Figure 18-1: Coastal Ocean Colour of Australian Waters: key activities and infrastructure

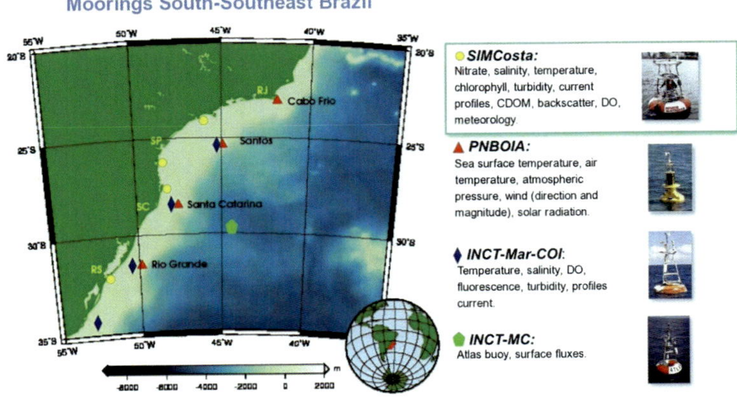

Figure 19-1: SiMCosta buoys along the Brazilian South-Southestern coast and other complementary arrays – PNBOIA, INCT-Mar-COI and INCT-MC

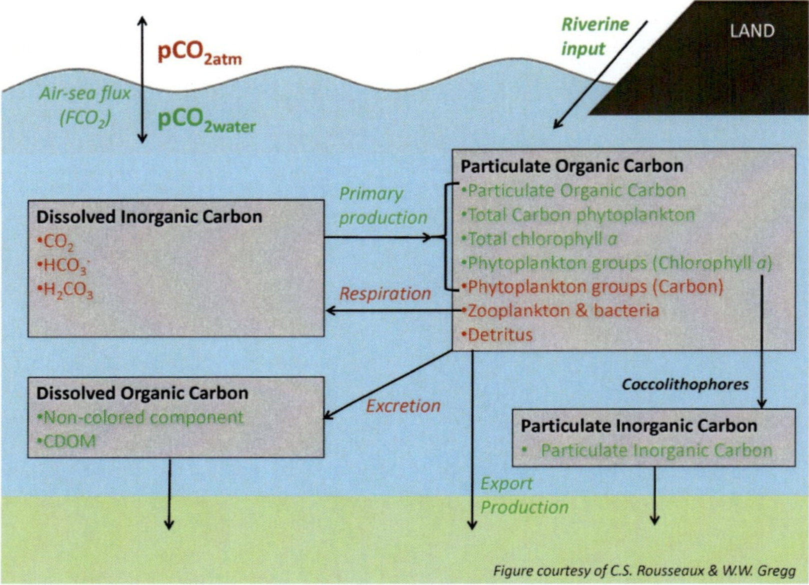

Figure 21-1: Details of biologically mediated carbon fluxes in the ocean; the dependencies of the components on other aspects of the carbon cycle are not shown. Green illustrates those for which satellite-based methods have been proposed in peer-reviewed literature and red those that are not. (courtesy of C.S. Rousseaux and W.W. Gregg)

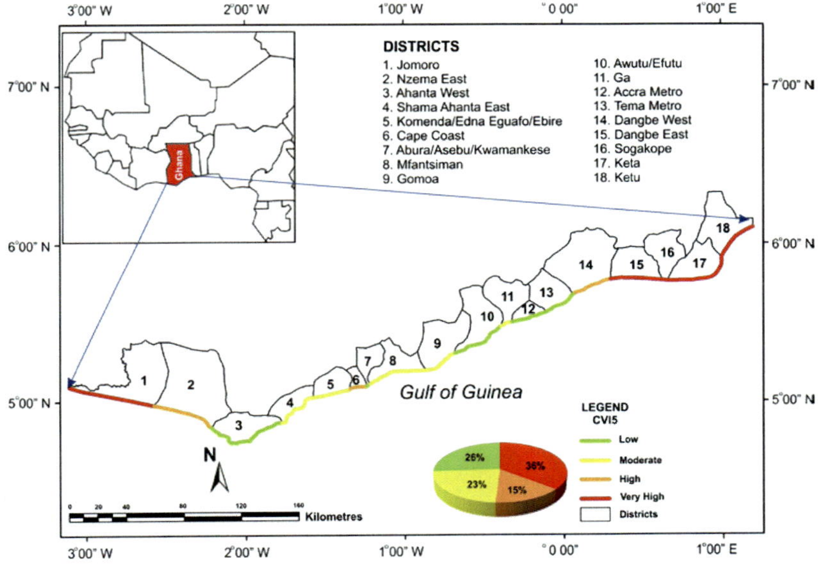

Figure 24-1: Erosion risk map of the Ghanaian coast

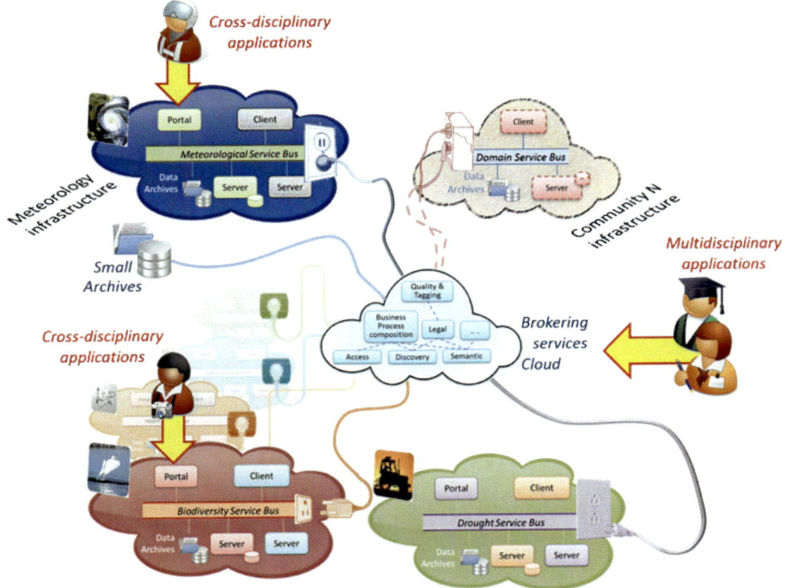

Figure 26-1: The Brokering approach

PART VI

OCEAN CLIMATE AND CARBON

Chapter Twenty

Ocean Remote Sensing and Global Climate Change

Carlos Garcia-Soto and José C. Báez

Introduction

In this chapter, we examine satellite measurements of global warming and of two associated major physical impacts upon the ocean: melting of Arctic sea ice and global sea-level rise. Sea-ice melting in the Arctic has been analysed by oceanographers since 1978 using passive microwave sensors and other types of satellite sensors. Mean sea-level increase has been measured using altimetry, and now gravity satellite sensors (GRACE), since 1993. We therefore focus on changes that have taken place during the last few decades, associated with the rise of global temperature due largely to increasing concentrations of greenhouse gases. This has been termed contemporary climate change. The chapter updates a previous report (Garcia-Soto *et al.*, 2012) and, as before, does not provide an exhaustive review, but rather gives the reader an insight into some key satellite observations of ocean climate change.

Changes in Sea Surface Temperature

Figure 20-1 A shows the time series of Global Land-Ocean Temperature Anomaly (°C) over the last century. The data set has been compiled and analysed by the NASA Goddard Institute of Space Studies (GISS) using SST measurements from satellites and buoys, measurements from Antarctic research stations, and Air Temperature from near 1,000 meteorological stations (Hansen *et al.*, 2010; NASA, 2013).

Overall, Figure 20-1 A shows that the global temperature has increased by 0.85°C since 1900, and that about 60% of that increase (~0.5°C) has taken place since 1980. This recent warming (1980–2010) gives a linear mean rate of ~0.17°C per decade, a warming twice as rapid as the secular mean (~0.08°C per decade; 1900–2010). At a more regional scale, a similar warming rate has been observed, for example, in the North Atlantic (Garcia-Soto *et al.*, 2002; Garcia-Soto and Pingree, 2012). The figure also highlights the overriding importance of the long-term temperature trend over the natural inter-annual fluctuations. The last year with temperatures cooler than the mean of the base period (1951–1980) was 1976. Despite the year-to-year natural variability, nine of the 10 warmest years on record have taken place since the year 2000, as usually indicated in the GISS reports. The warmest year since 1880 occurred in 2010, while the second warmest was recorded in 2005. In considering the rank between years, we need to take into account the fact that the uncertainty of the measurements (0.05°C; Hansen *et al.*, 2010) can be larger than the difference of adjacent global anomalies.

The global coverage of satellites allows us to elaborate global maps of temperature anomalies (Hansen *et al.*, 2010), which have shown that the largest positive anomalies are taking place in the Arctic. In this region, there is a positive climate feedback. The loss of ice, due to global warming, reduces the light reflected thus increasing the absorption of solar heat. Pole ward transport of heat towards the Arctic through the atmosphere and the ocean (Garcia-Soto *et al.*, 2002; Garcia-Soto and Pingree, 2012) can also play a contributing role. The global maps show that the record warming of 2010 took place during a transition from El Niño to La Niña in the Pacific. Therefore, 2010 was a remarkable year, not only because the highest annual increase of global temperature (since 1880) was observed, but also because that coincided with a major factor of natural variability (ENSO) being in a cooling mode. We note that, according to NASA (2010), El Niño of 1998 (El Niño of the Century) contributed to a global temperature increase of +0.2°C.

Changes in sea-ice extent

Figure 20-1 B shows the evolution of the extent of Arctic Perennial Sea Ice from 1979 to the present. The data set has been elaborated by the National Snow and Ice Data Centre, NSIDC, (Fetterer *et al.*, 2002) using passive microwave satellite data from the sensors SMMR (1979–1987) and SSM/I (1987-present). This type of satellite data allows for all-weather monitoring of sea ice, regardless of the time of day and cloud coverage conditions.

Overall, Figure 20-1 B shows a drastic reduction of the extent of Arctic perennial sea ice, decreased in the period 1979–2012 at a mean rate of 0.94 million km^2 or -11.5% extent per decade. This mean rate assumes a linear decrease for the full analysed period but the melting process has been accelerating in recent years. The minimum sea-ice extent of 3.63 million km^2 occurred in September 2012. This is a reduction of ~45% with respect to the mean of the 30-year base period (1981–2010, 6.52 million km^2). Since the year 2000, a historical minimum has been measured four times (2002, 2005, 2007 and 2012) and the last six years (2007–2012) have witnessed the six lowest minima since satellite measurements began (1979), as noted in the NSDIC reports (NSDIC, 2013a). Recent satellite studies (Comiso, 2012) have also revealed that the extent of the multiyear component of Arctic sea ice (>2 years) is decreasing at an even faster rate (~15% per decade) making the cover of perennial sea ice more vulnerable to global warming. Other signals of climate change in the Arctic sea ice include, for example, the timing and duration of the spring melting season that progressively starts earlier and lasts longer, increasing six days per decade (Serreze *et al.*, 2007). The thickness of Arctic sea ice has also decreased 25% per decade (from 3.64m in 1980 to 1.89m in 2008) as shown by combined records of submarines and data of the ICESat altimetry satellite (2003) (Kwok and Rothrock, 2009).

Ocean remote sensing reveals the effect of global warming on the perennial sea ice of the Arctic, but also on the ice shelves of Antarctica. An ice shelf is a floating platform attached to the coast, with a thickness between 50 and 600m, and which usually originates from the seaward flow of ice from land glaciers. In Antarctica, there are still 15 major ice shelves (Scambos *et al.*, 2007) and some of them are disintegrating, especially in the Antarctic Peninsula. The most prominent break-up of an ice shelf took place in 2002 when a large area of the ice shelf Larsen B collapsed, resulting in a loss of 3,200 km^2 in only 35 days (Scambos *et al.*, 2004). It was the most pronounced ice-shelf breaking in three decades. Satellite images from Synthetic Aperture Radar sensors have been used to monitor the shrinking size of the ice shelf over the years, and satellite images from the MODIS colour sensor have shown, for example, the presence of melting ponds (proposed as a disintegration mechanism) shortly before the collapse of the ice shelf Larsen B in March 2002. A summary of the retreat of different ice shelves in the Antarctic Peninsula during the period 1980–2008 is shown in NSIDC (2013b), including the last disintegration of the Wilkins ice shelf in 2008 (Scambos *et al.*, 2009). The disintegration of ice shelves is most prominent in Antarctica, but it has also been reported in the northern hemisphere including, for example, the fractures of the Ward Hunt shelf in Canada in 2002, 2008 and 2010 (NSDIC, 2013b).

Melting of the ice shelves and of the Arctic sea-ice do not have a significant direct effect on the rise of sea level (NSDIC, 2013c) since both components of the Cryosphere are already in the ocean. However, the disintegration of ice shelves significantly increases the speed of the glaciers that feed them, and melting of glacier ice is a major contributor to sea level rise. Following the 2002 disintegration of the Larsen B ice shelf, the associated glaciers of Antarctic Peninsula increased their speed up to eight times over the following 18 months (Scambos *et al* 2004). A comparable mechanism has been reported for glaciers in Greenland (Joughin *et al.*, 2004).

Changes in sea level

Figure 20-1 C shows the increase in Global Sea Level from 1993 to the present. The data base has been produced by AVISO (CLS/CNES/LEGOS) processing measurements from the altimeters Topex-Poseidon (1992–2005), Jason-1 (2002–2013) and Jason-2 (since 2008). Thanks to the global coverage of the altimeters, it is possible to measure the sea-level change not only at the coast, as before with tidal stations, but over the whole ocean (Cazenave and Nerem, 2004).

Of importance in Figure 20-1 C is the linear regression line that provides a rise of global mean sea level of 3.2 mm per year, and which accounts for 98% of the variability of the data (r^2: 0.98). The error of the slope given by AVISO after analysing the uncertainty of each altimetry correction is approximately 0.6 mm per year which indicates a value of global mean sea level rise since 1993 between 2.6 and 3.8 mm per year. There are several causes behind sea-level rise, and the budget of the different contributing factors changes with time (Church *et al.*, 2011; Steffen *et al.*, 2010; Pfeffer, 2011; Cazenave and Llovel, 2010; Cazenave and Remy, 2011). The major factors increasing global sea level during the altimetry era (1993–2010) have been, in order of importance: (i) ocean warming; (ii) melting of glaciers; (iii) melting of the Greenland and Antarctic ice sheets (including the glaciers of these regions); and (iv) the contribution of land water (AVISO, 2013). Ocean warming expands the volume of the oceans and the remaining three factors change the ocean water mass. The contribution of thermal expansion during 1993–2010 has been estimated at 1 mm per year, approximately 30% of the total sea level rise for the period. The contribution of the glaciers (excluding Greenland and Antarctica) has also been calculated as 1 mm per year. The mean contribution of the Greenland and Antarctic ice-sheets has been calculated as ~0.5 mm per year, about ~20% of the total. And the contribution of land water for the altimetry period has been estimated at about 0.4 mm per year. The budget is considered nearly closed given the uncertainties of the contributions (AVISO, 2013).

Some additional aspects of the sea-level budget should be considered. The different authors strongly emphasise the fact that the contribution of ice sheets to the total sea-level budget has significantly increased from an initial ~15% during the period 1993-2003 to the present decade (since 2003). This contribution is permanently being examined and refined. The contribution of land water to sea level also has a strong inter-decadal variability associated with natural climate oscillations such as ENSO (Nerem *et al.*, 2010; Cazenave *et al.*, 2012). This fact results, in association with changes of land water storage, in positive or negative inputs (increases or decreases of global sea level), depending on the decade. Finally, the Antarctic and Greenland ice sheets have contributed over the 1993–2008 period nearly 45% of sea level rise resulting from all the world glaciers, leaving the mountain glaciers or glaciers of low latitudes with as much as 55% of the total (Church *et al.*, 2011). This highlights the enormous relevance of the glaciers of low latitudes despite the fact that they represent only ~4% of the total land ice area (Meier *et al.*, 2007). To calculate the sea-level budget, researchers use altimetry data (measuring the total sea level, i.e., the steric plus the mass components), but also satellite gravity measurements (GRACE; measuring the mass variation and the GIA), temperature/salinity profiles (measuring *in situ* the steric contributions of temperature and salinity), and tidal stations (measuring the total sea level plus ground movements) among other methods.

The global nature of satellite altimetry allows us to draw global maps of sea level tendency that have shown that sea-level rise is not uniform (Lombard *et al.*, 2005, 2009). There are regions with pronounced increases while others show sea-level decreases. Analysis of the last 20 years of data reveals a strong regional signal in the Pacific Ocean, with the western region showing increases up to three times larger than the global mean and the eastern Pacific region showing negative values (Garcia-Soto *et al.*, 2012). The authors also analysed the mean sea-level trends of different sea basins and show large departures from the linear trend, for example in the Mediterranean Sea and the Black Sea, two semi-enclosed seas (Garcia-Soto *et al.*, 2012). The major reason for regional differences at a global scale is the non-uniform thermal expansion, associated largely to natural climate modes like El Niño (Lombard *et al.*, 2005; Meyssignac *et al.*, 2012). El Niño/La Niña cycles are driven by weakening and strengthening trade winds, making the associated changes of sea level a response to wind forcing (e.g., Merrifield and Maltrud, 2011). Other factors, such as changes in ocean circulation and the melting of the glaciers and ice sheets, are also known to result in regional differences of sea level (Mitrovica *et al.*, 2009; Willis and Church, 2012).

Anthropogenic effects versus natural variability

In order to quantify the anthropogenic contribution to the observed climate change, the influence of the natural climate variability must also be considered. El Niño or ENSO is probably the most important factor of natural climate variability at a global scale, and its influence on the Pacific large-scale sea-level patterns and on global warming has already been mentioned. Sea ice in the Antarctic region is known to be redistributed regionally also following El Niño/La Niña cycles (Rind *et al*, 2001; Kwok and Comiso, 2002). The importance of natural climate variability can also be observed in the Arctic where scientists have used satellite data to show an increase in the amount of melted sea ice in response to the positive phase of the Arctic Oscillation (Rigor *et al.*, 2002; Rigor and Wallace, 2004). Changes in the North Atlantic Oscillation, another atmospheric teleconnection, also result in significant changes in SST and sea-level anomaly in the North Atlantic Ocean (Garcia-Soto and Pingree, 2012). Separating anthropogenic from natural factors of climate change is not always straightforward, since the former can also influence the latter. For example, scientists have predicted a significant increase in the Central-Pacific-El Niño with respect to the Eastern-Pacific-El Niño under scenarios of global warming due to a flattening of the thermocline of the Equatorial Pacific (Yeh *et al.*, 2009). Similarly, the factors of natural climate variability have the potential to drive the natural system to a point where recovery is no longer possible under the present scenario of persistent global warming, as believed for example for the sea ice of the Arctic region (Lindsay and Zhang, 2005).

Links to Societal Benefits

In addition to the investigations of the natural and anthropogenic factors of climate change, current research is focusing increasingly on the study of climate change impacts. Direct impacts of global warming include not only the melting of Arctic sea ice and sea-level rise, but also coral reef mortality and the pole ward displacement in the distribution of fishery species. In the Arctic, sea-ice retreat is affecting the population of polar bears and other marine mammals. The increasing concentration of carbon dioxide itself is increasing the acidification of the ocean with damaging effects on marine calcifying organisms. Some direct consequences of sea-level rise on the coast include the loss of wetlands, beach erosion, storm surges, periodic flooding and salinisation of estuaries. Sea-level rise has created, according to the United Nations, a new type of refugee, the Climate Change refugee (UNEP, 2005), referring to Pacific island villagers that are relocated.

Over the land, the impacts of global warming include changes in precipitation patterns, decreases in biodiversity, and changes in the distribution of some disease vectors (IPCC, 2007). Africa is one of the most vulnerable continents as observed in the global maps of temperature anomalies (e.g., Garcia-Soto *et al.*, 2012). Agricultural production is projected to be greatly compromised by climate change which will decrease the area suitable for agriculture and the length of the growing seasons, increasing malnutrition.

International Cooperation

The major element of international collaboration in this domain of *Satellite Oceanography* is the IPCC, which will publish in 2014 his Fifth Assessment Report (AR5). This international effort has included more than 830 lead authors and review editors, with writing teams from 85 countries. Working Group 1 of this international panel assesses the physical scientific aspects of the climate change, including, among others topics, observed changes in air and ocean temperatures, ice sheets and glaciers, ocean sea level; satellite data and other data; and the causes and attribution of climate change; in a similar line to the review presented here. Working Group 2 assesses the vulnerability to climate change of the natural and socio-economic systems, and Working Group 3 assesses the options to limit or reduce the emissions of greenhouse gases and to remove the greenhouse gases from the atmosphere. We hope that the compromise arising from this international report on climate change observation, impacts, and particularly mitigation will allow us to confront this global challenge to our Blue Planet.

Conclusion

Global Land-Ocean Temperature Anomaly (°C), Arctic Sea-Ice extent (millions km^2) and Global Sea Level Change (mm) have been examined using combined satellite and *in situ* temperature data (1880–2012), data from satellite microwave sensors (1979–2012) and satellite altimetry data (1993–2012). Global temperature has increased near 0.85°C since 1900 and at about 60% of that increase (~0.5°C) has taken place since 1980. This recent warming (1980–2010) gives a linear mean rate of ~0.17°C per decade, a warming twice as rapid as the secular mean (~0.08°C per decade; 1900–2010). Satellite microwave observations of the perennial sea ice in the Arctic show a drastic reduction during the period 1979–2012 at a mean rate of 0.94 million km^2 or -11.5% extent per decade. The melting process has been accelerating in the recent years. The minimum sea-ice extent occurred in September 2012 (3.63 million km^2). This is a reduction of ~45% with respect to the mean of the 30-year base period. Altimetry measurements from Topex-Poseidon (1992–2005), Jason-1 (2002–2013) and Jason-2 (since 2008) show a rise of global mean sea level of 3.2 mm per year. The linear regression line accounts for 98% of the variability of the data (r^2: 0.98). The observations are analysed in the context of the recent scientific studies on Ocean Remote Sensing and Global Climate Change.

CONTEMPORARY CLIMATE CHANGE

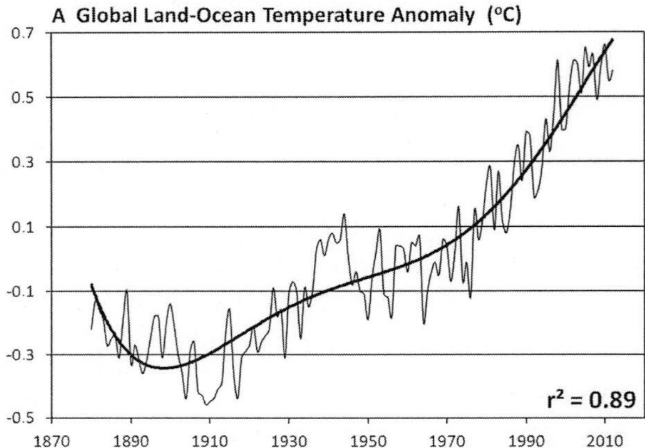

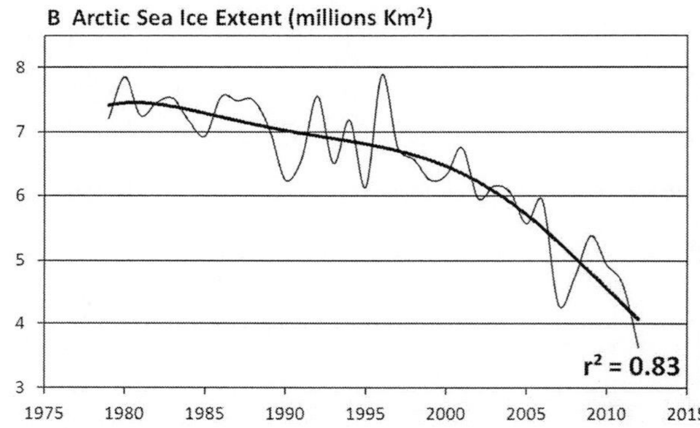

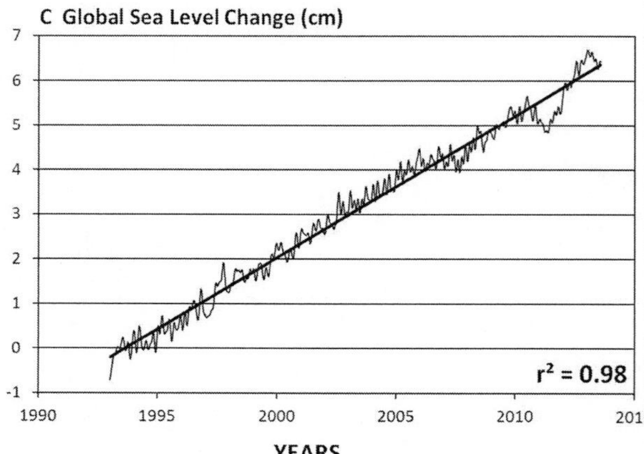

Figure 20-1: Time series of (A) Global Land-Ocean Temperature Anomaly ($^{\circ}$ C), (B) Arctic Sea-Ice extent (millions km^2; perennial sea ice extent) and (C) Global Sea Level Change (mm). The figures have been elaborated using combined satellite and in-situ data compiled by NASA-GISS (annual mean, 1880-2012), data from satellite microwave sensors from NSDIC (September values, 1979-2012) and altimetry data from AVISO (10 days resolution, 1993-2012). In order to highlight the long-term trends each figure includes a tendency line, which explains respectively 89%, 83% and 98% of the variability of the data. The tendency line of frames (a) and (b) are polynomial tendency lines of order 5. Polynomial tendency lines of higher order (6) are known to highlight additional shorter period variability (Garcia-Soto *et al.*, 2012)

CHAPTER TWENTY-ONE

CARBON OBSERVATIONS OF OCEANS AND COASTAL WATERS FROM SPACE

SHUBHA SATHYENDRANATH,
PRAKASH CHAUHAN, WATSON GREGG,
NICOLAS HOEPFFNER, JOJI ISHIZAKA,
JOHNNY JOHANNESSEN, MILTON KAMPEL,
TIIT KUTSER, TREVOR PLATT,
JOO-HYUNG RYU, DIANE E. WICKLAND
AND MARK DOWELL

Introduction

The GEO Carbon Strategy Report, published in June 2010 by the GEO Carbon Community of Practice (Ciais *et al.*, 2010), was a timely update to the Integrated Global Carbon Observations Theme report developed through the IGOS partnership in 2004–2005, and reflected the significant advances in science and capabilities since then. This new report identified the need for, and possible approach to, the implementation of an Integrated Global Carbon Observing (IGCO) system to address the three components of the carbon cycle (atmosphere, land, and ocean) and their interactions, and underlined the fundamental need for systematic global observations.

Recognising the need for a coordinated approach to ensure sustained observations of the global carbon cycle, CEOS also undertook a study. The resulting report (CEOS, 2014) presents the CEOS strategy for the planning and provision of space-based observations of the carbon cycle and its components in support of the various scientific and societal needs anticipated for carbon-related information. This report responds to the needs expressed in the GEO Carbon Strategy (Ciais *et al.*, 2010) and the ambitions therein for the realisation of an IGCO – with a specific focus on the satellite observations and the efforts of space agencies to provide them.

Here we reiterate and summarise the material for the Ocean domain from the CEOS Report, which is relevant to the Climate and Carbon sub-task of the GEO Blue Planet initiative. This chapter addresses requirements for the global oceans, regional seas, and coastal waters; specific issues addressing inland water bodies are dealt with in a separate chapter in the current monograph (see Chapter 6).

Background

Oceans constitute some of Earth's greatest reservoirs of the various forms of carbon: organic and inorganic, particulate and dissolved. It is estimated that the pool of carbon in the oceans is 50 times greater than that in the atmosphere; the flux of carbon through the ocean is much greater than that attributed to burning fossil fuels; and the atmospheric exchange of carbon with the ocean is larger than that with the land (Stewart, 2005). Primary production in the ocean is responsible for converting 50 Gt of carbon *per annum* into organic material (commensurate with terrestrial primary production), and a fraction of the produced material is exported to the deep ocean through sinking particles, leading to its sequestration from contact with the atmosphere. Ocean circulation transports carbon-rich waters from the surface into the deep ocean. The difference in partial pressure of carbon dioxide between the surface ocean and the atmosphere leads to exchanges of carbon between the two domains. Globally, the net exchange of carbon dioxide across the ocean-atmosphere interface has been such that some 25% of anthropogenic carbon-dioxide emitted into the atmosphere now resides in the oceans: without this uptake, the accumulation of anthropogenic carbon dioxide in the atmosphere would have been that much greater than it is today. But, over the years, the cumulative dissolution of atmospheric carbon dioxide into the oceans has modified the buffering capacity of oceans; this evolving role of the oceans has to be taken into account in planning for a carbon-neutral planet.

Planetary carbon fluxes, and the role of the oceans in them, cannot be discussed without considering the oceanic heat budget and air-sea fluxes of heat and momentum: the solubility of carbon dioxide in seawater changes inversely with temperature, and the distribution of temperature and salinity in the surface and near-surface layers of the ocean determine the total alkalinity in these waters. The air-sea exchange of the gas is determined by the air-sea difference in partial pressure of carbon dioxide and processes at the air-sea interface related to sea state, often parameterised as a function of wind speed. The physical and chemical processes that transport carbon in dissolved form from the surface to the interior are often referred to as the solubility pump and the biological processes that transport carbon (mostly in particulate form) to the deep ocean are referred to as the biological pump. Together, they create a complex picture, many details of which are yet to be clarified. The biologically-mediated carbon cycle in the ocean is presented schematically in Figure 21-1.

Although it is well recognised that ocean acidification due to dissolution of anthropogenic carbon dioxide is a serious threat to marine calcifying organisms, including corals, shell fish and many types of phytoplankton and zooplankton, the long-term impact of acidification on marine biodiversity has yet to be determined; it will depend, among other factors, on the ability of the organisms to adapt to the change. Changes in temperature, circulation, and stratification are modifying the distribution of many organisms, and further effects are anticipated in the future. Stratification determines the light available in the surface mixed layer for phytoplankton growth and, at the same time, absorption of light energy by phytoplankton modifies the heat content, creating feedback mechanisms between biological and physical processes in the surface layer, and hence in the carbon cycle. The implications for sustainable management of living resources of the sea are yet to be determined. Many geo-engineering schemes to sequester anthropogenic carbon involve perturbation of the pelagic ecosystem (for example, through iron enrichment or pumping of nutrients to the surface ocean from the deep). Before any such schemes can be considered seriously, it is necessary to understand the flow of carbon through the ecosystem, and the natural variability of this flow. Only then will it be possible to evaluate potential adverse effects on the ecosystem, as well as the magnitude of any potential sequestration. Thus, knowledge of air-sea interactions; mesoscale (order 100 km) to sub-mesoscale (1–5 km) dynamics in the ocean associated with the presence of eddies, meandering fronts and upwelling zones (Kudryavtsev *et al.*, 2012); and their interaction with, and influence on, the biogeochemical state of the ocean (Godø *et al.*, 2012) must be advanced.

Specific considerations for the Coastal Systems

Apart from the relevance of carbon pools and fluxes within the oceans in the context of climate change, it is also important to recognise the role of coastal ecosystems and near-shore habitats in carbon biogeochemistry. The coastal zone also represents a large reservoir of particulate organic carbon, resulting from local high productivity rates, as well as large inputs of land-derived organic material via river runoff. Upwelling of cold nutrient-rich waters, a typical phenomenon regularly observed in both *in situ* and satellite data (e.g., Kowalewski and Ostrowski, 2005; Kozlov *et al.*, 2012), leads to enhanced primary production. Approximately 30% of the oceanic primary production occurs in the coastal zone, which covers roughly 8% of the global ocean surface. Along the coastal fringe, sea grasses, seaweeds, benthic micro algae, rooted aquatic macrophytes such as mangroves, and coral reefs are major primary producers in the shallow environment ecosystems with very high rates of annual net productivity. Sea grass meadows occupy less than 0.2% of the global ocean area, but are estimated to contribute roughly 10% of the annual organic carbon burial in the oceans. The fate of the carbon fixed by these components varies with physical, chemical, and biological processes. The fixed carbon may be lost to sediments through burial or recycled within the system, consumed by herbivores, consumed within the detritus food web through microbial breakdown, or transported offshore by tides and currents as particulate or dissolved organic and inorganic carbon (continental shelf pump), and eventually become sequestered for several hundred years in the open ocean below the permanent pycnocline.

Although long considered a net source of carbon to the atmosphere, coastal waters can turn into a net carbon sink under increasing atmospheric CO_2. These fluxes are subject to large variability given that coastal zones are among the most dynamic, rapidly changing, and most vulnerable environments on earth. The coral reef ecosystems are particularly vulnerable to increased carbon dioxide concentration, which results in coral bleaching and loss of productivity. Understanding the fates of carbon sources and sinks within the coastal ecosystem, especially in the tropics, is important to establish the global carbon budget and to inform carbon-cycle models.

The role of satellites in monitoring carbon cycles and pools in the oceans and coastal water bodies

Carbon dioxide is reactive in the ocean, and biological processes add to the possible pathways for carbon; carbon occurs in many particulate and dissolved forms. The oceans, which are dynamic and subject to variability on multiple time and space scales, are perennially under-sampled. *In situ* observations based on ships and buoys cannot, by themselves, provide the coverage necessary to detect potential changes superimposed on long-term variability. Satellites provide repetitive observations with global coverage, serving as an extrapolation and integration tool, filling gaps in *in situ* observations, especially in the horizontal plane at the surface. Ironically, it is sometimes easier to detect particular carbon fluxes, such as primary production, in the ocean, by combining remotely-sensed data with auxiliary information, than it is to detect some of the carbon pools themselves. Furthermore, satellites are able to inform us on many physical factors that influence the transport of carbon through the ocean, and the flux of carbon at the air-sea interface.

The contributions that satellites can make to monitoring pools and fluxes of carbon in the ocean are summarised schematically in green text in Figure 21-1. Note that many of the important components of the surface ocean carbon system are now routinely observed from satellites, a credit to the international space agencies that have provided and managed operational remote-sensing platforms, as well as to the scientific community for innovation in finding methods to convert the raw radiances directly observed into scientifically useful geophysical products. It is, moreover, reassuring that the sensors required are consistent with the Essential Climate Variables identified by GCOS. However, it is important to note that until the satellite products have been refined to the point where they can be expressed in carbon units, or can contribute to measurements of carbon fluxes, such an observing system does not constitute a carbon observation system. This is particularly true of ocean colour, where many novel and emerging products need to be exploited and developed further. To be useful, space observations must be capable of evaluating and reducing uncertainties in the estimates of these pools and fluxes, and to be able to monitor, on a routine basis, small changes in these fluxes. Because the natural system is highly variable, even over decadal time scales, observations have to be sustained in a systematic manner over a very long time, to allow anthropogenic trends to be isolated from natural variability. At the same time, the observation system has to be multidisciplinary and integrated, and capable of identifying potential changes to the marine ecosystem and its services, in addition to changes in the capacity of the oceans to remove anthropogenic carbon dioxide and other greenhouse gases from the atmosphere.

Remote-sensing observations of the ocean carbon cycle are restricted to the surface layer of the ocean. A broad representation of the surface ocean carbon cycle (Figure 21-1) illustrates the main components, and also shows what is uniquely observed from space by a single sensor class of observations (e.g., ocean colour radiometry) or indirectly estimated with more than one sensor class (e.g., thermal infrared and optical scanners, radar altimeters and scatterometers).

Present gaps and long-term sustainability of satellite observations

There are major gaps in the observing systems. The entire dissolved inorganic carbon component is not observed, despite the fact that it is the largest pool, and one that is undergoing considerable change. Unfortunately, this component has insufficient electromagnetic signal to be detectable from space using current technology. But the on-going SST measurements and the new salinity measurements have the potential to contribute to inferences of changes in this pool.

In summary, the most pressing needs for remote sensing in support of ocean carbon science are:

1. Continuity of the current observational methodologies;
2. New missions with improved capabilities; and
3. New observations of atmospheric pCO_2.

Since no remote-sensing methods to measure atmospheric pCO_2 currently exist, it is important to have adequate *in situ* coverage of atmospheric pCO_2 measurements; these will not only complement satellite measurements of related variables, they will also serve as validation points for future methods that may be developed to remotely sense atmospheric pCO_2.

High on the list of aspects of the carbon cycle that are poorly observed and quantified are the pools and processes in coastal regions. For example, ocean-colour products continue to suffer from degraded quality in these optically-complex areas. New observations are needed, along with advances in engineering: broad spectral ranges, high spectral resolutions, and a capability to observe the smaller scale biological and chemical dynamics that characterise these regions.

The complementarity of satellite and *in situ* datasets for global ocean carbon cycle analysis

While most studies on the need for satellites do not mention the importance of supplementary *in situ* data, the GEO Carbon Strategy Report (Ciais *et al.*, 2010) highlights this need. Initiatives such as Argo, OceanSITES, AERONET, ChloroGIN, and bio-Argo float programmes are welcome developments in this context (also note the *in situ* component of the OCR-VC initiative of IOCCG and CEOS). The usefulness of long-term sustained *in situ* observations increases with the value of the data record, as has been well demonstrated, for example, in the case of the Continuous Plankton Recorder programme. Not only is the value of satellite observations enhanced when coupled with *in situ* observations for testing and validation of the methods, but in many cases the satellite-derived parameters cannot be inferred without the *in situ* measurements. Along with the need for *in situ* data is the need for archival and distribution by international data centres. *In situ* data are also required for extrapolation in the vertical dimension, for example, to link surface chlorophyll with vertical structure in chlorophyll profiles (e.g., Platt *et al.*, 2008). *In situ* observations are essential to establish the indirect methods for detecting carbon pools, and to ensure that methods stay robust over time, in a changing ocean.

Conclusion

Climate change has the potential to modify many chemical and physical processes in the ocean, and hence the capacity of the oceans to take up anthropogenic carbon dioxide from the atmosphere. For example, changes to stratification and circulation would impact cycling of dissolved inorganic carbon (labile and refractive components), and warmer ocean temperatures and increasing partial pressure of carbon dioxide in the oceans would affect the further uptake of carbon dioxide from the atmosphere.

When considering the relevance of the ocean carbon cycle in the context of climate change, it is not sufficient to examine only how the carbon cycle through the oceans dictates the accumulation of carbon dioxide in the atmosphere and hence the strength of the green-house effect. It is also important to recognise that it is the flow of carbon through the marine food chain that sustains the marine ecosystem and marine resources, particularly fisheries and seafood. It is not yet known how climate change might modify marine ecosystem services, including marine primary production, and food from the sea, that are taken for granted.

The availability of sustained observations is necessary to face upcoming challenges related to the detection of long-term trends and cycles of variability. A continuous dataset also supports the development and application of data assimilation and reanalyses. Models constrained by satellite data will allow us to examine how the ocean responds to climate variability, to identify potential long-term trends, and to improve carbon estimates in regions of low and biased sampling, such as the Southern Ocean. Satellite observations serve as inputs to models designed to study the flow of carbon through the oceans, and to validate model outputs. This broader role of remote sensing in contributing to our understanding of the ocean carbon cycle should not be overlooked in designing the satellite component of an ocean carbon observing system. Programmes such as the Climate Change Initiative of ESA, are now highlighting the need to bring many missions together in a consistent fashion to produce long time series of essential climate variables, and to link data with models to achieve an integrated view of the Earth's climate system.

Ideally, the ocean-carbon observation system would be an integrated system, incorporating both *in situ* and satellite observations, rather than treating individual observing elements as stand-alone tools. The need for *in situ* observations is often undervalued, but together with satellite data the combination can lead to major improvements, especially when combined with modelling and data assimilation efforts, potentially producing the types of scientific advances in carbon-cycle science that are needed.

PART VII

DEVELOPING CAPACITY AND SOCIETAL AWARENESS

Chapter Twenty-Two

Towards Sustained Ocean Observations in Developing Countries

Sophie Seeyave, Shubha Sathyendranath, Trevor Platt and Victoria Cheung

Introduction

The ocean has many beneficial impacts on society, for example, provision of food, renewable and non-renewable energy, recreation and means of economical transport. On the other hand, disasters can be caused by natural phenomena relating to extreme weather (e.g., storm surges), climate change (e.g., sea level rise), or geophysical events (e.g., earthquakes, tsunamis), which in turn are intrinsically linked to the ocean. The exploitation of marine resources can also lead to major disasters with impacts on human health, tourism and recreation (e.g., HABs and oil spills). With ocean and society so closely linked, it is vitally important that we observe the oceans on a global scale, and in an integrated manner, to better manage the exploitation of our marine resources, and to improve prediction and mitigation of disasters. Oceanic processes do not respect political boundaries, therefore it is the responsibility of all nations to ensure the sustainable use and responsible management of the ocean, and international coordination of such efforts is essential. Currently, the resources (financial, human and material) required for sustained ocean observations are concentrated mostly in developed countries, and vast areas of the world ocean remain largely unobserved, most notably in the Southern Hemisphere. To redress this imbalance, there is a growing need to develop expertise in ocean observations in many coastal nations worldwide, particularly, but not exclusively, in the Southern Hemisphere. This chapter outlines what has been achieved to date by individuals and nations, the value of international coordination and the role international organisations can play. In addition it provides some examples from the POGO, an organisation that has a strong mandate for marine capacity building.

Capacity building:
From individual to country-wide efforts

Past examples have demonstrated that, in terms of capacity building, individuals can make a difference. One such example is Dr N.K. Panikkar (1913–1977) from India who: created a network of fisheries research institutes in India and the Indian Committee on Ocean Research; led India's participation in the International Indian Ocean Expedition (IIOE); laid the foundations for the National Institute of Oceanography in Goa; founded the Indian Ocean Biological Centre in Cochin; and was the father of TEMA, the Training, Education and Mutual Assistance Programme of the IOC. In the words of Gotthilf Hempel, Dr Panikkar was "one of the world's most successful capacity builders in marine sciences" (Hempel, 1998). His major message was to "invest primarily in young people rather than in equipment and vessels which tend to rust so quickly in tropical waters".

A number of developed countries have invested much effort and funding in developing oceanographic research in emerging countries and now have strong capacity building programmes. One such example is Norway, in particular through its Nansen Centres established in countries such as South Africa, India, Russia and China. Their aim is to "serve the society through advancing knowledge on the behaviour of the marine environment and climate system in the spirit of Fridtjof Nansen" (Dr Nansen was a Norwegian scientist, polar explorer, diplomat, humanitarian and Nobel Peace Prize winner). The Nansen Scientific Society that advocates "Knowledge without Borders" funds *education and research within global environment and climate problems, including their impacts on society*, through a fellowship programme, research projects and expeditions, scientific meetings and conferences. In addition, the research vessel RV Dr Fridtjof Nansen has provided support for fisheries surveys off the coasts of more than 60 developing countries. All these programmes are inspired by a great man and his passion for advancing knowledge, and through these, Norway is making significant contributions to capacity building in marine science around the world.

The need for international coordination

Because the ocean transcends national boundaries, and because it is so vast and challenging to access, oceanographic research requires international collaboration. Furthermore, capacity building requires at the very least two-way interaction between developing and developed countries, for example through exchange of personnel and expertise. Such exchanges can be, and are, facilitated by international organisations such as the POGO, IOC and the Scientific Committee on Oceanic Research (SCOR). Capacity building can be most effective through the participation of developing countries in research projects run in collaboration with developed countries. Ultimately, if capacity building is truly successful, developing nations will become more integrated and active in international networks, such as those named above.

One of the first major multi-national collaborative efforts in marine capacity building was the IIOE, which took place from 1959 to 1965 under the auspices of SCOR and the IOC, to describe and understand the basic features of the Indian Ocean. This project came about due to the realisation that the Indian Ocean had remained relatively unexplored compared to the Atlantic and the Pacific Oceans. It was important to address the fundamental oceanographic problems linked to monsoonal cycles experienced by the northern Indian Ocean; to determine the chemical characteristics of the water column and the abundance and distribution of food resources such as fish productivity; and to understand the geology of the Indian Ocean through sea bed mapping and sampling. Studying these features was considered important not only to the global community of researchers, but also to the large population that lives in the countries around the Indian Ocean (Laughton, 2004).

The Government of India was an enthusiastic participant in this expedition. The IIOE involved forty-six research vessels under fourteen different flags, and covered all parts of the Indian Ocean. As the IIOE approached its concluding phase, the Indian government decided that the Indians who participated in the expedition needed to have an institution where they could build on the oceanographic research skills they had acquired during the expedition. Under the leadership of N.K. Panikkar, the National Institute of Oceanography (NIO) was subsequently born. Today, NIO is one of the major oceanographic institutions in the world. In his account of the expedition, Anthony Laughton concluded that "As a result of the Expedition, the Indian Ocean became one of the most systematically studied oceans of the world and revealed a large variety of geological features and processes not seen in the Atlantic or the Pacific. The

Expedition's success must be considered to be one of the major achievements of SCOR and IOC." (Laughton, 2004).

Another historical example of a collaborative project involving developing countries was the Global Atmosphere Research Programme (GARP), a joint undertaking of the WMO and ICSU. The First GARP Global Experiment involved three cruises conducted by the Instituto Oceanografico of the University of São Paulo (IO-USP), Brazil, in collaboration with US and Italian partners. Today, IO-USP is contributing towards the establishment of a Climate Observing System in the South Atlantic, using moored buoys such as the PIRATA and ATLAS-B arrays, which are part of GOOS (see Chapter 15).

International organisations such as IOC, SCOR and POGO work together to ensure that capacity building is carried out in an integrated manner, avoiding duplication of effort. For example, POGO and SCOR collaborate to deliver a Visiting Fellowship programme, and the IOC's International Ocean Data and Information Exchange (IODE) provides training in data management at the Centre of Excellence in Observational Oceanography that is run by POGO in partnership with the Nippon Foundation. The three organisations meet regularly to discuss a common strategy, identify needs and plan new activities, such as, for example, setting up the Ocean Summer Schools web portal in 2011 (http://www. oceansummerschools.org).

GEO also has a strong mandate for capacity building, with particular emphasis on societal benefits. Examples of GEO initiatives include ChloroGIN (see Chapter 12) and SAFARI. ChloroGIN aims to promote *in situ* measurement of chlorophyll in combination with satellite-derived estimates and associated products to facilitate assessment of marine ecosystems for the benefit of society. With support from GEO and GOOS, it developed from a regional network that was established in Latin America, itself the outcome of a training course that took place in Chile. ChloroGIN is recognised as a pilot project of GOOS. The principal objective of SAFARI is to coordinate, at the international level, applications of remotely-sensed Earth Observation data to the societal benefit areas of fisheries and aquaculture. Both ChloroGIN and SAFARI were Tasks within the first phase of implementation of GEO (2005-2011) and are now included in the "Oceans and Society: Blue Planet" Task. Each of these programmes are effectively international networks of researchers that emphasise the need for developed and developing countries to work together towards a common goal.

The POGO example

POGO (http://ocean-partners.org), created in 1999 by directors and leaders of major oceanographic institutions around the world, is a forum to promote global oceanography, particularly the implementation of an international and integrated global ocean observing system. From its inception, POGO recognised that capacity building was a prerequisite to attain its objective of creating a global ocean observing system. In 2001, in its "São Paulo Declaration", the Participants of POGO called upon the world leaders of government, industry, science, and education to use their influence and resources, within their own countries and through relations with others, to devote necessary attention to extending ocean observing systems in the Southern Hemisphere, as a minimum requirement towards implementing an integrated strategy for observing the global oceans. This Declaration was used by the University of Concepción (Chile) to justify why Chile was uniquely positioned to make a valuable and unique contribution to the field; by the Japan Agency for Marine-Earth Science and Technology to justify a circumpolar cruise of their research vessel Mirai in the Southern Hemisphere; by the Royal Netherlands Institute for Sea Research (NIOZ) to justify a project for three long-term time-series stations (two in the Indian Ocean); and by the NOAA Laboratory, AOML, to initiate a new XBT line from Cape Town to Buenos Aires.

The first step in capacity building for POGO was the establishment of its Visiting Fellowship programme. This programme was initially run in collaboration with IOC and SCOR, but has continued since 2006 as a joint POGO-SCOR scheme. Under this programme, scientists from developing countries can spend up to three months training at a major oceanographic institution. By 2013, 150 young scientists from over 30 countries had been trained under this scheme (see Figure 22-1). A variation on this programme is the POGO-AMT Fellowship, which allows one scientist annually to participate in the major international and interdisciplinary Atlantic Meridional Transect cruise led by Plymouth Marine Laboratory and the National Oceanography Centre in the UK. Participation in such a prestigious programme is an opportunity that is rarely given to young scientists from developing countries. In 2005, POGO partnered with the Nippon Foundation (NF) to set up new initiatives in capacity building. The NF was established in 1962 as a non-profit philanthropic organisation, active both in Japan and abroad, with a strong interest in the marine environment. This collaboration started with the NF-POGO Visiting Professorship Programme under which scientists visit a developing country to conduct training in ocean observations. This included training in India, Sri Lanka, Viet Nam, Fiji, Tunisia and Brazil. In 2008, this programme metamorphosed into the NF-POGO Centre of Excellence in Observational Oceanography, now the flagship of POGO training. Under this programme, ten young scientists, mostly from developing countries, are supported each year to study for ten months in an intensive programme related to ocean observations. The first phase of the Centre of Excellence (until 2012) was at the Bermuda Institute of Ocean Sciences (BIOS) and the second phase is at the Alfred Wegener Institute in Germany (from 2013). Meanwhile, POGO has continued the Visiting Professorship programme on a more modest scale, with its own funding, which so far has included teaching in Argentina, Namibia, Viet Nam, and Sri Lanka. POGO also supports African graduate students to study at the University of Cape Town, South Africa, and provides travel support for participants from developing countries to attend the Austral Summer Institute at the University of Concepcion, Chile.

This suite of training programmes is making a very significant contribution to reducing the deficit in trained observers of the ocean in developing countries. Under POGO capacity-building schemes, around 500 young scientists from over 60 countries have received advanced training. The massive over-subscription for POGO training schemes provides ample proof that the effort is responding to a genuine need. Networking is seen by GEO, POGO and NF as a very important part of the training process. Former scholars or alumni of NF-POGO training become members of the rapidly-developing NF-POGO Alumni Network for Oceans (NANO, http://www.nf-pogo-alumni.org), which has a newsletter and several regional projects that are carried out jointly by the alumni. The topics are of local, societal relevance (e.g., HABs and coastal pollution) and aim to enhance international cooperation and ocean observations in developing nations. POGO and NF also recognise the value of keeping sight of their former trainees, to continue to nurture them as much as possible and to monitor their progress as they become the oceanographers and leaders of tomorrow.

Conclusion

After half a century, Dr Panikkar's position that we should invest in people first has been substantiated: with the right people in the right places, other assets follow. This has been the case for India and Brazil, two examples, among many, of success stories in oceanography. However, such developments do not happen overnight; capacity building can be extremely challenging, albeit rewarding. It is worth emphasising the need for long-term investment in human capacity development. As part of the process, it is important to follow up on the training, to monitor progress post-training and to integrate the trainees in international networks of oceanography. International organisations can play a vital role in providing and coordinating capacity building efforts, and they should work together to maximise their effectiveness.

When planning future capacity-building initiatives, it is important to identify and fill gaps in the ocean observing system and to target societal needs. Equally, we should find ways to assess the impacts of our capacity building programmes and identify ways in which they can be improved. We must identify centres in developing countries that have shown drive and initiative in providing training and education at regional levels, and promote their activities. We must identify and nurture the future leaders who will pass on the knowledge gained to their peers and to the next generations and become the "Panikkars" of tomorrow. We should encourage north-south partnerships and initiatives where both parties benefit from the effort, and encourage south-south partnerships to disseminate regional expertise. Finally, we must encourage developing countries to take the lead in deciding on the types of capacity building that best serve their needs.

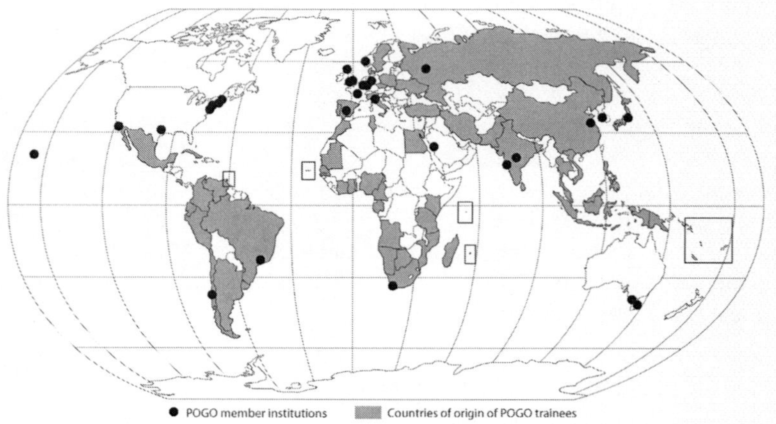

Figure 22-1: Map showing the locations of POGO member institutes and countries of origin of POGO trainees

Chapter Twenty-Three

Satellite Ocean Colour Radiometry and the Role of the International Ocean Colour Coordinating Group (IOCCG)

Venetia Stuart

Introduction

The IOCCG was established in 1996 to promote communication and co-operation between the various space agencies and the scientific ocean-colour user community as well as to build capacity in both developing and developed countries. Functionally, the group operates as an Affiliated Programme of SCOR and as an Associate Member of CEOS. Representatives from major international space agencies as well as members from the scientific research community serve on the IOCCG Committee according to a rotating schedule (see http://ioccg.org/about/ members.html for current list of members). Several space agencies, as well as other oceanographic organisations, contribute financially to the IOCCG and help to carry out the decisions endorsed by the group, while the scientific members address current research issues and make recommendations.

Links to Societal Benefits

One of the main products of remote sensing of ocean colour is the concentration of chlorophyll-a, which is used as an index of phytoplankton biomass. Since phytoplankton standing stocks are responsible for transforming carbon dioxide into organic carbon, they play a major role in global climate change. Satellite ocean-colour radiometry is thus an important tool which can be used to examine biological and biogeochemical processes of marine ecosystems on synoptic scales, with a temporal resolution of 1–2 days. With the advent of more complex and sophisticated ocean-colour sensors, it is now possible to retrieve a wide range of products from space, with an extensive range of applications. In addition to the concentration of chlorophyll-a, it is also possible to retrieve products such as the concentration of total suspended matter, coloured dissolved organic material, particulate organic and inorganic carbon, primary production, aerosol properties, and phytoplankton functional types, to name but a few. Ocean colour radiometry has many important applications including understanding the role of the oceans in the global carbon cycle, quantifying the impacts of climate variability and change, assessing ocean ecosystem health and productivity (including monitoring harmful algal blooms), and managing ocean ecosystems and resources. For example, ocean colour radiometry data, together with other satellite data, can be used to help classify habitats of species at risk (or commercially exploited species), to better elucidate their distribution and migration patterns, and understand their response to external forces such as climate change and fishing pressure. Furthermore, since fish stocks often congregate around certain oceanographic features that can be recognised using satellite ocean colour and SST data (e.g., fronts, eddies, upwelling areas) this information can be used to help increase the efficiency of fish harvesting, resulting in a reduced search time and a corresponding increase in fuel savings. The benefits of ocean colour technology to society are wide and varied, which is why ocean colour observations are a key component of Earth observation programmes of many major space agencies.

A wide range of initiatives, ranging from scientific and technical issues to capacity-building efforts, are addressed by the IOCCG Committee each year, as discussed below. Furthermore, the IOCCG is taking a key leadership role in helping to ensure the continuity and consistency of the ocean-colour data stream through the CEOS Ocean Colour Radiometry-Virtual Constellation (OCR-VC).

IOCCG Scientific Working Groups

An important focus of the IOCCG is the establishment of specialised scientific working groups to investigate various aspects of ocean-colour technology and its applications. Generally, the end product of these working groups is the publication of an IOCCG monograph on the topic, which has evolved into the highly-acclaimed "IOCCG Report Series". This report series has earned the IOCCG wide international recognition for scientific excellence, and is used extensively by ocean colour scientists and managers, both for research as well as for policy-and decision-making. To date, thirteen IOCCG monographs have been published by scientific working groups (see http://ioccg.org/reports_ioccg.html), with a number of other working groups nearing completion. These reports are widely cited in the scientific literature and are seen as the definitive work on the topic.

The IOCCG report series started with a report edited by Prof. André Morel on *Minimum Requirements for an Operational Ocean-Colour Sensor for the Open Ocean* (IOCCG Report 1, 1998). Information contained in this report has been critical for the design of operational ocean colour sensors for many years. Satellite instruments and missions are currently evolving beyond traditional measurements, which has led to the publication of the latest IOCCG report on this topic (IOCCG Report 13, 2012) entitled *Mission Requirements for Future Ocean-Colour Sensors*. Other IOCCG reports have addressed remote sensing of ocean colour in coastal and estuarine waters (IOCCG Report 3, 2000), highlighted key issues for binning and merging ocean-colour data (IOCCG Reports 4 and 6, 2004; 2007), provided a comprehensive evaluation of societal benefits and applications of ocean-colour radiometry, including fisheries (IOCCG Reports 7 and 8, 2008; 2009), and provided a broad summary of ocean colour algorithms and atmospheric correction algorithms (IOCCG Reports 5 and 10, 2006; 2010). In addition, IOCCG working groups have addressed the requirement for complementary ocean colour missions (IOCCG Report 2, 1999), ocean colour remote sensing from a geostationary platform (IOCCG Report 12, 2012), partitioning the ocean into ecological provinces (IOCCG Report 9, 2009), and examining the feasibility of equipping Argo floats with bio-optical sensors to validate ocean colour data (IOCCG Report 11, 2011). Current IOCCG working groups focus on other topical issues, including remote sensing of phytoplankton functional types, calibration of ocean colour sensors, ocean colour remote sensing in Polar Seas, and remote sensing of HABs (see http://ioccg.org/groups_ioccg.html for list of current working groups). The highly anticipated reports from these working groups should be published in the near future.

Continuity and Consistency
of the Ocean Colour Data Stream

Another important issue being addressed by the IOCCG is the continuity and consistency of ocean colour radiometry datasets. An uninterrupted, well-calibrated global ocean colour time series is a very valuable observational resource for climate studies. However, the data requirements for ocean colour applications are so diverse that data from a single satellite sensor cannot meet all ocean colour needs, so data from multiple missions must be used to augment data availability. Several space agencies are focussing on generating a continuous time series of observations by merging multiple satellite data sets to support research needs as well as applications. This requires consistency in the calibration of instruments and algorithms used, as well as in the data products, and is only achievable through international cooperation among the space agencies. These issues are currently being addressed by the IOCCG through the CEOS OCR-VC, comprised of a number of international space agencies that are working together to add value to individual missions (e.g., through cross-calibration, improved validation, and data merging). The OCR-VC aims at producing sustained ocean-colour radiometry data records of well calibrated and validated satellite data from measurements obtained across multiple satellites.

Within the OCR-VC framework, the *International Network for Sensor Inter-comparison and Uncertainty Assessment for Ocean Colour Radiometry* (INSITU-OCR) initiative aims at integrating and rationalising inter-agency efforts on satellite sensor inter-comparisons and uncertainty assessment for remote sensing products, with particular emphasis on requirements addressing the generation of ocean colour Essential Climate Variables (ECV) as proposed by GCOS. Under the guidance of the IOCCG, representatives of space agencies and institutions supporting INSITU-OCR agreed on a series of recommendations that are critical to ensure high accuracy and consistency among products from present and future ocean-colour missions. These recommendations can be viewed at: http://www.ioccg.org/groups/INSITU-OCR_White-Paper.pdf.

A related initiative is the recent establishment of an IOCCG Task Force on *Ocean Colour ECV Assessment* which is carrying out an independent evaluation of ocean-colour ECV products produced by different agencies, to be used for climate-related studies. This working group will compare ocean colour ECV products from the same sensor produced by different methods, as well as products from different sensors produced using the same methodology. Furthermore, comparisons with independent ground-truthing as well as comparison to target requirements, as defined by GCOS, will help to establish confidence limits for a long and coherent time series of global ocean-colour ECV products.

International Ocean Colour Science (IOCS) Meetings

Recognising the importance of maintaining consultation and interaction with the broader ocean-colour user community, the IOCCG resolved to conduct a series of international ocean-colour science and consultation meetings in different parts of the world – provisionally every two years – to provide a forum for discussion of various topics related to ocean-colour radiometry, thereby engaging ocean-colour scientists and space agency representatives from around the world. The first IOCS meeting, co-sponsored by NASA, EUMETSAT, ESA and CNES, took place in Germany (6–8 May 2013), and was attended by 244 participants including representatives from all the major international space agencies as well as members of the global ocean-colour research community. The format of the meeting included invited keynote speakers, information talks by space agency representatives, and splinter sessions on issues related to the establishment of ocean colour Climate Data Records, with a significant amount of time allowed for discussions, as well as poster sessions to highlight the latest research in the field of ocean colour radiometry (see http://iocs.ioccg.org/). Splinter sessions participants came up with a series of recommendations which will be carried forward to space agencies, as well as explored further through a number of follow-up workshops. These IOCS meetings will ultimately lead to the strengthening of the international ocean colour scientific community by engaging a wide range of users from around the world.

Education and Outreach

Another strong focus of the IOCCG is training and capacity building. Since its inception, the IOCCG has sponsored and conducted over twenty training courses on the theory and applications of ocean-colour data, thus helping to augment the ocean-colour user community, particularly in developing countries. A recent focus of the IOCCG has been on high-level training of bright young researchers, through the "IOCCG Summer Lecture Series". The first of these training courses entitled *"Frontiers in Ocean Optics and Ocean Colour Science,"* took place in Villefranche-sur-Mer, France (2–14 July 2012), and was dedicated to advanced training in the fundamentals of ocean optics, bio-optics and ocean colour, focussing specifically on current critical issues of concern in ocean colour science. A total of 17 students from 12 different countries attended the course, and received in-depth lectures and practical sessions from a number of distinguished ocean colour specialists. Interactive discussion sessions between the students and lecturers formed a significant portion of the course, and students were exposed to a wide array of topics, providing an invaluable insight into research questions and approaches that could be applied to their current research as well as future careers. A second Summer Lecture Series is scheduled to take place from 21 July to 2 August 2014, thus helping to create a high-level research community with the necessary background and skills to address current critical issues, especially those related to global climate variability and change.

Conclusion

The various IOCCG meetings and training initiatives help to bring together researchers from around the world, thus promoting and enhancing international collaboration in all areas of ocean colour research, as well as highlighting requirements and applications of ocean colour data. This in turn helps to strengthen the global ocean-colour user community and will lead to an increased demand for ocean colour data products and services around the world.

The informal cooperation and international coordination amongst IOCCG member Agencies helps to promote exchange of data and the development of compatible data products for current and planned ocean colour missions, to optimise societal benefit. Furthermore, through its report series, the IOCCG addresses issues of common interest in ocean colour radiometry providing expert advice to the global user community.

CHAPTER TWENTY-FOUR

DEVELOPING COASTAL RESEARCH IN GHANA

AUGUSTUS VOGEL

Introduction

Of the approximately 196 countries in the world, 152 have a coastline. Given that the ocean provides important economic benefits such as transport, tourism, and access to raw resources, ocean and coastal science can play a very tangible role in supporting the development of these coastal nations. However, many of these nations are developing countries and do not have the resources or expertise to fully utilise the science. Capacity building in coastal and ocean research is therefore, almost by definition, a key element in the development and economic stability of coastal nations.

Capacity building is not, however, the core mission of many organisations that do research. In fact it could be argued that the two don't seem to mix very well when performed in developing countries. Programmes that focus on capacity building often seem (according to the author) to be very weak on actual science (evidence of success being measured by how many people are "trained", not by what those trained people actually do) and programmes focused on science often have capacity building tacked on as a very junior partner.

It was this context that confronted the United States Office of Naval Research (ONR) when it decided that it might be interested in supporting coastal geoscience research in Africa. ONR has a mandate to fund research of benefit to the United States Navy, and that research does not necessarily have to occur within the United States. It funds a variety of topics (which can be seen at http://www.onr.navy.mil), including the understanding of the ocean and coastal environment. The concept in the case of Africa was based on the question of how ONR might fund a project, and in so doing, help establish a research effort that would continue beyond the initial ONR funding. The benefits would be mutual: ONR would see an increase in scientific articles about ocean and coastal environments, particularly in a part of the world without extensive referencing in academic journals, and the partner country would benefit from the increase in research, the importance of which was noted in the first paragraph.

However, ONR falls decidedly in the camp of funding research, not capacity building, and was not prepared to specifically allocate funds with the goal of simply training African researchers. To be fair, students are often funded by ONR and within the context of their education in support of the project, some "capacity building" does occur. It just doesn't happen in the development assistance sense. The challenge was therefore how to work with an African partner in such a way that the needs of both sides were met and quality research was performed.

Finding a Partner

The first critical element was to find a partner with whom ONR could work. This process was initiated by funding a number of African researchers to visit with ONR representatives at the IEEE International Geoscience and Remote Sensing Symposium (IGARSS) in Barcelona Spain in 2007. These researchers were identified through discussions with other United States and international agencies and perusal of websites, reports from international meetings, and academic journals. A decision was made to focus on West and Central Africa. Although the ONR team had members that had previously worked in Africa, they did not have specific experience with ocean researchers from the continent. ONR was looking for researchers that had relevant skills and experience, and therefore spent a significant amount of time looking through various resumes. There were three characteristics for which they were specifically searching:

1. **Patience:** Given that ONR was not familiar with the researchers selected, a slower process was initiated that would provide sufficient time to develop an acceptable project. ONR did not want to simply solicit, in an impersonal way, a series of white papers that may or may not have worked out.
2. **Energy and Creativity:** It was perhaps unreasonable that ONR wanted to fund research in Africa without actively making capacity building a critical element. Either way it was left to the funded researcher to resolve this issue through the sourcing of local expertise and contacts in other organisations. Given that the intention was for the funded research to act as seed funding, it was

important to find individuals who also had the desire to leverage the support to continue the research.

3. **Willingness to be a True Partner:** It can be argued that a lot of research funded in Africa is not particularly balanced, given that much funding comes from international sources and the funders decide how the funds can be spent. Corruption has also had a detrimental effect on many programmes and has instigated efforts to track activities and spending that have very little to do with trust. ONR wanted to work with an African partner to produce strong science, but expected that local issues would be solved on the side without encumbering the project. In this way ONR could interact with the African partner in a way that was indistinguishable from how it collaborates with other funded scientists.

These three characteristics are not necessarily obvious in a resume, but would rather require a certain amount of interpretation matched with face-to-face discussion and further interaction to demonstrate ability for the two sides to work together. At the end of the IGARRS, ONR suggested that a series of technical workshops be developed in cooperation with the African invitees. In those technical programmes ONR-funded scientists would spend a week with African scientists, and in so doing develop a scientific project of mutual interest.

First Steps

The only workshop that went forward was at the University of Ghana at the now Department of Marine and Fisheries Science (MFS). This self-selection process left ONR with only one option, but this was the partnership for which it was searching. The Ghanaians were clear that their goals were not to benefit United States Naval Science, but that they hoped to use the research to increase their ability to support the various coastal development issues that their country faces. These include high levels of coastal erosion, rapid population growth along the coast, and looming port developments driven by the nascent oil industry. Given that this would be achieved by publishing fundamental research on the African coast, both sides were satisfied.

A small team of ONR-funded scientists visited Ghana in April of 2008, where they had a one-week workshop on coastal modelling and data collection in sandy environments. A field trip introduced the ONR scientists to the Ghanaian coast, and the last day of the programme involved the team laying out a three year proposal that would be of interest to ONR and of use to Ghana. The Ghanaians also immediately started the process of bringing together local stakeholders. By later that year ONR had sent the initial funding.

The Research Effort

The overall goal of the project has been to develop a cause-and-effect understanding of the Ghanaian coastline. To achieve this, work commenced on three principal areas of research:

1. **Modelling:** The Ghanaian team, in collaboration with the University of New Hampshire, deployed a wave buoy to better model the environment within the Bight of Benin. They also sent a student to UNESCO's International Water Institute to learn how to use Delft3D, an established modelling system for coastal environments. Unfortunately, deployment of the wave buoy has been problematic because of larceny attempts by the large artisanal fishing fleet off Ghana. Another mooring is being purchased to re-deploy the buoy into waters that are less than ideal for regional modelling but which represent a reduced risk from the canoes.
2. **Remote Sensing:** Collaboration with the United States Geological Service has helped Ghana to start using aerial and satellite imagery to monitor historical change on their coastline. It too has faced challenges such as weak internet service and expensive high resolution imagery in Ghana. However, MFS now has a regular group of students working on satellite imagery who are increasingly finding free or low-cost sources of imagery.
3. *In situ* **Measurement:** Using reference markers, a GPS-RTK, and a small echo sounder, the Ghanaian team has developed a risk map for the entire coastline of Ghana (Figure 24-1). This has been the area in which the Ghanaian research has advanced the most, and progress has allowed them to finally create a true baseline for Ghana, against which the country can accurately quantify erosion and changes to the coastline.

Diversification of the Partnership

As identified by the Ghanaians, a primary goal was to make the research relevant to the development of their country. To achieve this, outreach was performed to establish collaboration with civilian government offices, civil society, and the Ghanaian Navy. This effort was naturally organised and led by the Ghanaian researchers. Initially there were hurdles, however the department has been able to establish a more complementary role in the country. They can for example, offer the opportunity of graduate degrees in an area which would have normally required an expensive stay in Europe.

Growing the student body in MFS has also been a challenge. Young students are not necessarily equipped with the skills to perform some of this research. Programming skills in Matlab and R for example, are not normally taught in their curriculum but are basic requirements for modern-day science. ONR has also been able to help somewhat with this issue, by using reservists to introduce students to basic programming skills. Combined with an increase in the size of the department to capture returning Ghanaian researchers, the project has been able to bring in four PhD students, six Masters students, and seven Undergraduate students.

Finally, the Ghanaian group has been working to link their research with other research efforts, including those that are regional and continental. It is a way of leveraging and diversifying resources, but also expanding their profile. One example is installation of a DevCoCast receiving station as part of the EAMNet project, an EU funded programme that helps bring satellite products to Africa.

Making an Impact

It has been important to identify how the output can be used. A book on the research results was being published in 2013, establishing the basic research credentials of the effort. Partnerships within Ghana though, have been the key area for making the work relevant. MFS has performed a series of stake-holder meetings, and maintains collaborative partnerships with the Fisheries Ministry, the Ghana Maritime Authority, and the Survey Office. Further, they have signed a Memorandum of Understanding with the Ghana Navy to provide research and technical support for coastal bases and operations. These are not titanic shifts, but represent the work needed to create the long-term relationships that will eventually allow for fundamental successes. For those who have worked in African development, a recurring theme is that it is often easier to contract outside the country than to try and solve issues locally. Ghana for example has traditionally solved its erosion problems by contracting foreign companies to survey and "repair" their coast. By creating and supporting local research capability that can also graduate professionals, Ghana is changing this dynamic.

Conclusion

One important consequence of the programme is that the University of Ghana increased its competitiveness and won a five million euro grant from the European Development Fund to be a regional centre for maritime research. A sister centre exists in Mauritius, but to date West Africa had not been able to establish its own version. While this still does not qualify as local funding supporting local solutions, it does represent the success of a programme developing to the point that it can leverage current funds to receive new monies. Certainly ONR was pleased that its three year project was going to continue to grow.

Seeing the project as a success up to now, the following lessons were taken from the effort:

1. Emphasising the research over the capacity building is a reasonable strategy, but which requires full engagement with the African team to build a collaborative effort.
2. The two partnering sides do not need to have the same ultimate objectives, so long as the work itself can meet the objectives of both sides.
3. Motivated individuals on both sides are needed for the work to be successful. The outreach and diversification of the project for example, have been successful because they were led by the Ghanaian partner.

CHAPTER TWENTY-FIVE

CHALLENGES OF TRAINING AND CAPACITY DEVELOPMENT IN DATA AND MARINE INFORMATION MANAGEMENT IN THE XXIST CENTURY

ARIEL H. TROISI

Introduction

Data constitute the first step in a) creating and/or supporting knowledge, and b) developing information (i.e., captured data and knowledge), and both elements are necessary for informed decision making. Therefore data and information (D&I) management is a truly cross-cutting activity, and the marine D&I domain is clearly no exception. Training and capacity development are key to ensuring that all interested parties, regardless of their social and cultural backgrounds, can equally participate in data observation and research activities, and can thus effectively respond to the needs and requirements of their stakeholders.

Within the marine domain, the D&I community is at a cross-roads after facing several challenges in a rapidly changing environment, each of which has had a direct impact on the need for capacity building and development:

- The emergence and growth of well-funded data systems, linked with other data systems in different regions;
- Acceptance, adoption, and adaptation of best practices and standards;
- The need for interoperability;
- Creation of a quality-management framework;
- Dynamic changes in the target community, from single national entities responsible for ocean- related data management to many ocean research and observation "entities" (including universities,

research institutions, projects and even small groups of researchers);

- The evolution of data management from a technology-intensive activity involving heavy investment in very expensive data processing technology, to the development of PCs with computing power and ability to produce data and information for the whole citizen spectrum;
- The development of the Internet and World Wide Web.
- Data collected automatically (on-board, platform based), which is processed, stored and served without intervention of specialised data centres;
- Increasing data volumes and diversity, and the resulting problems related to integration, assimilation, and analysis for data product generation; and
- The social impact of science and knowledge evolution. There is an increasing demand to take into account where the investments in science are being placed, the emerging trends and the returns to society of those investments considering also social, workforce and economic indicators.

The role of data and information management in the advancement of ocean research and monitoring

Added value to the ocean research and observation community could thus consist of providing expertise linked to data and information management and related management plans. For the different elements of these processes (e.g., metadata schemes, and quality control), manuals, user guides, and technical procedures are indeed required, however these alone do not guarantee an appropriate transfer of knowledge. In a scenario of multiple and diverse data sources, with direct access to data sets, products, and information, physical (or electronic) documentation is still not enough. Capacity development efforts should also include outreach, public awareness, and communication strategies to, *inter-alia*: empower end-users to better understand and assess the data and information they are accessing and using; determine if the sources are authoritative; make informed decisions on its use, based on quality control, instruments, platforms, methodologies; and even be able to contact the data/information originator or curator.

As mentioned above, technical developments have led to an ever-increasing volume of data, as well as new data types and formats, that require advanced skills to process, analyse, and preserve it for future use.

Linked to this are the initiatives related to data citation and publication, through which data originators receive credit for their work; researchers make clear reference to the exact data sets used, ensuring the possibility of replication by others; and D&I managers have to deal both with preserving the "picture" of a moment, while tending to the continuous development and growth of databases and holdings.

Moreover, increased speed in communications and requirements for real-time and near-real-time data are just two of the challenges facing D&I managers. These same issues, however, have also become part of the daily life of forecasters, climatologists, and researchers, raising the need for D&I management training in academic and operational institutions. The transfer of knowledge has always been complex. Nonaka and Takeuchi (1995) stated that "...where the only certainty is uncertainty, the one sure source of lasting competitive advantage is knowledge." As proposed by Enkel (2002), the possibility of combining organisational forms such as "knowledge networks", communities of practice, and ICT tools such as portals, webinars, and teleconferences, might lead to suitable management systems to face the current and near future challenge of transferring implicit and explicit knowledge. Nevertheless, one of the critical factors in a virtual community's success is its members' motivation to actively participate in community knowledge generation and sharing of activities.

The marine D&I environment is not different from those in other disciplines; earlier studies have postulated the existence of a certain resistance in sharing knowledge in some environments (Ciborra and Patriota, 1998). Knowledge might not flow easily even when an organisation makes a concerted effort to facilitate knowledge exchange (Szulanski, 1996), or where the success of knowledge exchange depends on the organisational knowledge management system's social and technological attributes (Holthouse, 1998), and on organisational culture and climate (DeLong and Fehey, 2000). The benefits and value of D&I management are only realised, however, when project proposals and initiatives have sufficient time, human resources, infrastructure, and funding allocated to each component of D&I management: data policies that will be followed, provision for long-term stewardship of data, and D&I training and capacity-building.

Another key element in D&I management is the proper identification of shareholders and stakeholders, and addressing their requirements. Whilst a traditional perspective would consider a linear path from researchers to D&I managers to end users, resource managers, decision makers and policy makers, a more suitable approach considers them as cogs within a complex system. Consequently, training and capacity building/development in marine D&I have to be understood and considered in a holistic way, taking into account not only the shareholders but also the stakeholders. The aim should be achieving synergies that avoid duplicating or overlapping effort or, even worse, inadvertently generating voids or generating creative solutions for inexistent problems.

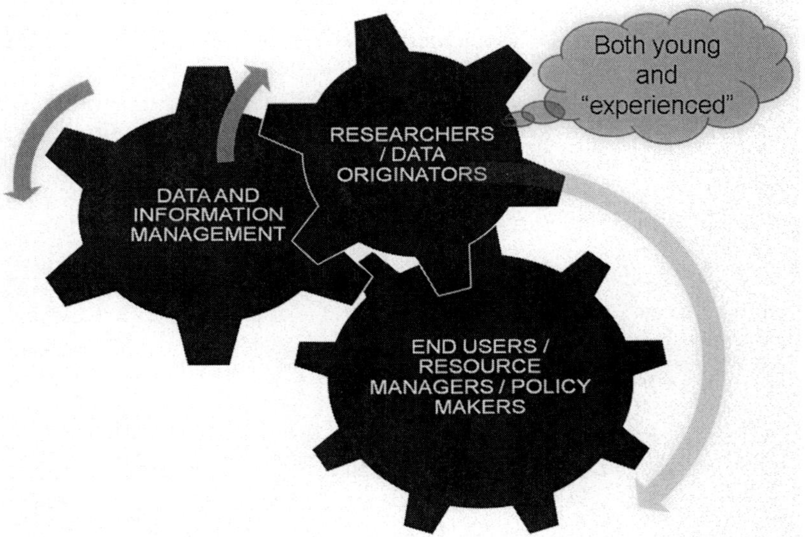

Figure 25-1: Relationship between Researchers, D&I managers and End Users

Innovative approaches to bridge the knowledge divide

Old structures and methods no longer serve today's user needs. Knowledge-based systems of training courses have been developed to synthesise the data resources available from operational programmes (including wind, wave, current, surface temperature, salinity, and sea-ice datasets), allowing the assembly, synthesis and display in single software platforms. Furthermore, these courses help demonstrate the enhanced value of multiple observing systems.

Although current information and communication technologies provide new ways to implement these activities, intensive efforts are still needed to bridge the digital divide. In addition, there are issues related to language barriers and increasing travel and organisational costs to set up face-to-face capacity building and development activities. The language issue is not a trivial one, even when highly technically qualified assets and trainers are identified; the lack of a common language (or limited language skills in such a common one) limits knowledge transfer efforts.

In this sense, working in a regional context provides, in many cases, the ability to surmount language problems and, at the same time, addresses common (regional) as well as individual (national) needs, aiding in the development of links and communities of practice. On the other hand, however, regional approaches might present additional difficulties, such as insufficient Internet bandwidth and computer infrastructure to support the training courses in certain locations.

An example of such a strategic approach is the Ocean Data and Information Network (ODIN) strategy adopted by the IODE programme of the IOC. Together with the ODIN strategy, IODE started the development of an Internet-based training tool supplied with suitable contents, initially called the IODE Resourcekit (1997), later renamed OceanTeacher (2001). IODE was the first programme in IOC to publish materials (the ResourceKit) on CD-ROM and use html for both content and user interface.

One of the core success elements of the ODIN strategy is the two-tier approach for product and service development: while ODIN projects deliver regional products and services (e.g., regional databases), each partner country also receives support to develop products and services that are specific to national or even local priorities and needs. This has led to a wealth of products developed at the national level, from research-oriented taxonomic databases to "what do I find on the beach?" brochures aimed at primary school children. This approach was found to maximise involvement and buy-in from the partner institutions and partner experts, thereby optimising the potential for long-term sustainability of the established infrastructure and expertise.

In a further step, IODE recently launched the OceanTeacher Global Academy which will take advantage of existing ICT tools to change training from a "north to south" culture to a north-south, south-south, and south-north model promoting collaboration and expertise exchange through internet-based technologies such as video conferencing and video streaming. Whereas training was traditionally based on experts from developed regions visiting and teaching students in developing countries, the OceanTeacher Global Academy will promote expertise available in many developing regions, contributing to the sustainable management of oceans and coastal areas worldwide.

Specifically, the OceanTeacher Global Academy will promote the establishment of Regional Training Centres that will: plan, organise, and implement training courses that are locally relevant and serve needs within their region; promote the use of local experts as lecturers and training assistants within the Regional Training Centres; and promote collaboration between the Regional Training Centres by enabling (through advanced information technology) lecturers from multiple regions to contribute lectures and further develop the OceanTeacher Learning Management System to cover multiple IOC (and associate) programmes.

Such initiatives will lead to several benefits, including an increase in the availability, involvement, and level of expertise of trainers; linking classrooms in geographically dispersed locations; alleviating the costs and other drawbacks of long-distance travelling by trainers/lecturers and trainees, contributing to an increased focus on local issues while keeping a global perspective; and last but not least, increasing self-driven capacity development, including local training expertise.

Adding to the complexity is the human resources challenge. Marine D&I management highlights the need for marine scientists knowledgeable in information, in communication technologies, and in resources coming from the ICT field, and also interested in understanding the intricacies of the different disciplines of Marine Sciences, each requiring specific skills and knowledge (e.g., physical data with its relatively low complexity and high volume; biological data with a much higher complexity and lower volume; marine chemistry and biochemistry, numerical modelling, remote sensing, etc.). The scarcity and value of such candidates often results in competition for them, leading to intensive staff rotation.

Finally, the aforementioned skills and knowledge must be combined with those in the Social Sciences. Unless this is achieved, at least at management and project leadership levels, the efforts will produce limited results, both in terms of scientific results and social impact of investments in science and science management. It is critical for those involved in marine D&I management to understand societal requirements and manage expectations, to envisage possible needs and scenarios and, of course, to acknowledge and bridge the communication gap between marine D&I curators and managers, end-users, and policy and decision makers.

Conclusion

In view of the rapid technological developments of the last decades, in particular the ICTs, the data and information community is certainly at a cross-road. This community thus needs to make critical decisions to ensure proper response to an ever-increasing demand from their end-users. In this sense, the knowledge management strategy and systems become critical as they constitute the fundamentals for the training and continuing capacity development of the limited human resources devoted to marine data and information management.

Besides considering purely technical issues, those involved in the development of such strategies and systems have to take into account the key role of outreach and public awareness activities in helping identify and target end-user audiences, responding to societal needs, and securing recognition and support of marine data and information management activities. Current technologies should be utilised to their fullest extent as they offer suitable capabilities to address the aforementioned issues with due and proper consideration to local and regional conditions and situations.

CHAPTER TWENTY-SIX

GEOWOW:
A FRAMEWORK FOR MULTI-DISCIPLINARY INTEROPERABILITY OF OCEAN DATA AND SYSTEMS

MASSIMO CRAGLIA AND STEFANO NATIVI

Introduction

The societal, environmental, and economic value of the oceans is enormous and is increasingly recognised as a major resource for humanity. However, the sustainable exploration, exploitation, and protection of ocean resources require a knowledge base and predictive capabilities, which are currently fragmented or not yet readily available.

The delivery of predictive capability will require strengthened collaboration in the collection of information from both *in situ* and remote-sensing ocean observation systems, as well as in research and innovation activities that use these ocean observation data to model ocean processes.

The European Union (EU) Expert Group on Marine Research Infrastructures (2013)[1] describes the current European marine and ocean observation landscape as:

- Densely populated with high quality projects, networks, initiatives, and sectoral visions which are, more often than not, disjointed, and where, without a unifying vision or framework, the *whole is significantly less than the sum of its parts;*
- Expensive and resource-intensive in terms of personnel time, energy, and ideas; the current cost of existing marine and ocean data collection is about €1.4 billion ($1.8 billion) *per annum;*

---

[1] http://ec.europa.eu/research/infrastructures/pdf/publication_b5_mri_report.pdf.

- Complex, where this complexity hinders the development of a coherent European (and/or transatlantic) capacity for marine observation.

Efforts to develop observation capability are being pioneered in the United States, for example, by the Ocean Observatories Initiative (OOI)[2], and by a variety of initiatives involving stakeholders from the EU and countries associated with the EU Research Framework Programme (for example, projects such as EMSO, EuroSITES, COOPEUS, ODIP), the EU maritime policy initiative EMODnet, that is facilitating access to these data, as well as initiatives from Canada (e.g., NEPTUNE).

Within Europe, MyOcean, the marine service of the Copernicus initiative, provides operational ocean forecasting on a global and regional scale, and delivers boundary conditions for more detailed coastal analyses. Another important initiative is SeaDataNet, which has developed an efficient, distributed Marine Data Management Infrastructure for the management of large and diverse sets of research data deriving from *in situ* and remote observation of the seas and oceans. The goal of the EU "marine knowledge 2020" programme is to deliver a multiple resolution seabed map of European waters by 2020 along with enhanced ocean forecasting capabilities. Notwithstanding these important initiatives, there is still much to tap fully into the potential the oceans offer to our economy and society.

The report *Science for an Ocean nation: update of the Ocean Research Priorities Plan*[3], released in February 2013 by the US Subcommittee on Ocean Science and Technology National Science and Technology Council, states that:

Progress has been made through a number of international entities, but connectivity and coordination is still needed in several areas:

- Integration in terms of data compatibility. The compatibility of datasets is a critical challenge;
- Connection of observations on the cryosphere, atmosphere, planetary boundary layers, land, and ocean in models along with ecosystems, organisms, and humans over different scales to better understand weather, climate, and ecosystem processes and services. This level of integration requires an interdisciplinary level of thinking beyond what exists today;

---

[2] http://ooi.oceanleadership.org/.
[3] http://www.whitehouse.gov/sites/default/files/microsites/ostp/ocean_research_plan_2013.pdf.

- Advocacy for more effective regional implementation of ocean policy objectives through internal collaboration;
- Establishment of international minimum standards for data collection to facilitate data sharing around the globe. Such sharing would require new capabilities to address gaps or differences in data collection, sharing, and interoperability of technologies, and permit integration of existing research into operational models; and
- Education and development of scientists, managers, and other experts capable of operating in this environment and able to integrate physical, ecological, mathematical, and social data into decision-making tools.

GEOSS[4] addresses some of these challenges, supporting the interoperability of multi-disciplinary information systems at the global level, with a particular focus on the oceans through the Blue Planet initiative launched in 2012.

The European Commission strongly supports the development of GEOSS and of Blue Planet through a number of dedicated research projects and initiatives. GEOWOW (http://www.geowow.eu) is one such EU-funded research project, promoting GEO interoperability across Weather, Ocean ecosystems, and Water run-off. GEOWOW has a strong connection with Blue Planet through one of its leading partners, the IOC, that is also one of the key leaders of Blue Planet. GEOWOW supports the IOC in developing the information systems and tools, including cloud-based tools, necessary to make a global assessment of the oceans as part of the Trans-Water Assessment Programme. It also develops further the brokering framework necessary to bridge gaps across the information systems and capacity of different disciplines. The following section discusses this particular contribution by GEOWOW to the interoperability of ocean information systems and data.

Interoperability

Interoperability was defined by IEEE as "the ability of making systems and organisations to work together (inter-operate)". While the term was initially defined for information technology or systems engineering services to allow for information exchange, a more effective definition must take into account social, political, and organisational factors that impact system to system performance (Slater, 2012).

Historically, interoperability has been pursued by adopting two complementary approaches: (a) pursuing standardisation (i.e., recognising

---

[4] http://www.earthobservations.org/geoss.shtml.

a common set of standards as a baseline to form a federation of systems) (Nativi *et al.*, 2006; Domenico *et al.*, 2006; Nativi *et al.*, 2013); (b) providing intermediation capabilities (i.e., developing brokering services to resolve the mismatches characterising complex technological environment, such as multi-disciplinary and global frameworks) (Heimbigner and McLeod, 1985). Traditionally, multi-disciplinary interoperability has been pursued on a case-by-case basis or by asking stakeholders (i.e., both users and resource providers) to utilise the large number of different protocols, standards, and agreements (also known as community-of-practice service buses) that characterise different disciplinary infrastructures (e.g., federations). Clearly, this represents a high entry barrier for developing cross-disciplinary scientists and applications.

For this reason, a new solution was proposed, first by a European FP7 project[5] (Santoro *et al.*, 2012; Nativi *et al.*, 2012) and then by a United States National Science Foundation initiative[6], namely: the Brokering approach, to provide intermediation capabilities in a transparent way.

The Brokering approach is based on the following principles (EarthCube, 2012) (Nativi *et al.*, 2013):

a. **Autonomy**: keep existing disciplinary infrastructures as autonomous as possible, not asking them to implement any "more general" service buses.
b. **Subsidiarity**: supplement, but not supplant, disciplinary infrastructure mandates and governance arrangements by interconnecting and mediating their service buses.
c. **Interconnection**: incrementally build on existing infrastructures and introduce distribution and mediation functionalities to interconnect the heterogeneous service buses characterising any specific domain or other infrastructure.
d. **Low entry barrier**: assure a low entry barrier, for both users and resource providers, of any disciplinary infrastructure.
e. **Flexibility**: be flexible enough to accommodate existing and future information systems and information technologies that will augment the service bus implemented by any discipline.
f. **Effectiveness**: address the full range of information exchange needs (discovery, access, semantics, workflow, etc.).

---

[5] http://www.eurogeoss.eu.
[6] http://www.nsf.gov/geo/earthcube/.

As depicted in Figure 26-1, the Brokering approach introduces a new middleware layer of service offerings: the Brokering framework, depicted in Figure 26-1 as a cloud. This can contain all the necessary existing (and new) components and services, including brokers to implement interoperability among present (and future) service buses of different disciplines. Therefore, a Broker may be defined as an intermediary middleware, dynamically implementing a many-to-many interconnection for a Client-Server framework. A Brokering framework is a third-party infrastructure which is not part of any disciplinary infrastructure, providing services at the multidisciplinary level.

Brokers realise the necessary mediation, adaptation, distribution, semantic mapping, and even quality checks required to address the complexity of a multi-disciplinary infrastructure built by interconnecting disciplinary infrastructures. Multidisciplinary users can access heterogeneous resources transparently, either by using a Community service bus (i.e., protocols and data models), or by directly accessing the Brokering service.

With respect to existing mediation solutions, a Broker presents the following advanced features:

a. Implements both proxy and gateway functionalities in a Client-Server framework.
b. Implements intermediation in a dynamic way –i.e., at runtime.
c. Allows new types of interoperability interfaces to be easily plugged in and configured.
d. Is neither a server-side nor a client-side component or service.

Several types of brokering services have been developed and tested (Vaccari *et al.*, 2012; Nativi *et al.*, 2012; Bigagli *et al.*, 2013), including:

- Discovery broker
- Access broker
- Semantic (discovery) broker
- Quality broker
- Policy broker
- Business process broker

Flexibility can be more easily achieved if the broker functions can be structured as modules in a Broker Framework. This is the approach that has been developed and demonstrated. The first three modules or broker types listed above will be discussed below; while the final three are still at differing stages of development, and will not be discussed in this chapter.

Discovery Broker

To enable discovery of multidisciplinary resources (i.e., not only data but also services, documents, models, etc.), a "discovery broker" must implement the "traditional" functionalities: discovery, messaging, and managing with additional capabilities (Nativi and Bigagli, 2009; Bigagli *et al.*, 2004):

a. **Extended messaging**: to provide support for asynchronous communication patterns, to accommodate lengthy operations, and to implement push-style interactions (e.g., distributed queries returning partial results as soon as they are available). Important issues to be addressed include: the retrieval of partial query results, query status checking, and query process interruption. To enable this, the following advanced features are enabled: incremental query, query feedbacks, query interruption.

b. **Enhanced distribution**: to allow communication between a client and multiple servers of the same kind, each presenting as if they were a unique instance (cardinality issue). Thus, the main tasks of an ideal "distribution component" are appropriate request routing and aggregation of responses.

c. **Mediation**: to allow communication between a client and a server of different type (heterogeneity issue). The main task of an ideal "mediation component" is to integrate heterogeneous server and client components by adapting their technological (protocol), logical (data model), and semantic (concepts and behaviour) models. In fact, dealing properly with heterogeneity, by mediating metadata profiles as well as through protocol bindings, is central to the successful deployment of multi-disciplinary discovery services.

d. **Paging and ranking**: to show distributed query results on multiple pages instead of just putting them all on one long page, a well-used solution to return and display a long list of matching results. Paging solutions must be coupled with a ranking strategy applying unified metrics. This is to return and display the "most relevant" results first.

Together, these capabilities allow a discovery broker to support asynchronous query distribution over heterogeneous (disciplinary) service buses. A discovery broker must implement both harvesting and query distribution approaches.

The Access Broker

An Access Broker facilitates the access and use of available resources, for example, the datasets discovered by using a discovery broker (Nativi *et al.*, 2013). This access broker provides two types of functionalities:

(a) transformation services to interconnect access clients and servers – a typical brokering task;
(b) supplementary services to complete access and allow use of the resources accessed. Examples of these capabilities are: Coordinate Reference System (CRS) transformation, resampling and sub-setting functionalities, and encoding format transformation.

Generally, an Access Broker implements the following features:

1. The distribution of a client access request, which may contain one or more datasets to be accessed, to the proper access server(s), providing consistent feedback. This entails:
   i. Mediation between the client request model and the server(s) model(s), in both directions.
   ii. Adaptation between the client request protocol and server(s) protocols, in both directions.

2. Allowing the user/client to access all the data requested according to a Common Grid Environment (CGE). CGE defines a common CRS, spatial and temporal resolution, spatial and temporal extent, and data encoding format.
3. Allowing the user to configure his/her strategy for supplementing access servers, by selecting their trusted capabilities (i.e., third-party services), as often, users require control of the transformations selected to supplement access servers and, sometimes, to use specific (disciplinary) capabilities. This is carried out by publishing a configuration interface to enable users to:
   i. choose a default third-party service to implement dedicated processing (e.g., CRS transformation);
   ii. upload a specific third-party service and set it as the default choice;
   iii. select the order of the processing flow (e.g., first CRS transformation, then spatial resampling, etc.).

The Semantic Discovery Broker

In a multidisciplinary environment, "traditional" discovery interfaces may be ineffective for users who are not familiar with the terminology in other disciplines. Semantic assets, like controlled vocabularies, thesauri, and ontologies are useful to lower this barrier (Nativi *et al.*, 2011; Nativi *et al.*, 2012; Nativi *et al.*, 2013). They relate generic terms and concepts with disciplinary specific terms and concepts, allowing non-expert users to formulate sophisticated query constraints. This is achieved by "expanding" users' terms (i.e., query clauses) with a set of disciplinary-specific terms which are semantically related to the users' base. This approach is called semantically augmented discovery, and can be implemented by a Semantic Discovery Broker. This approach is also used to address multi-lingual resources. Semantic Discovery Brokers generally implement two different augmented discovery styles: (i) automatic query expansion and (ii) user-assisted query expansion. In the first case, users' terms (i.e., keywords used as query clauses) are expanded automatically by the Broker. This is done by interrogating a set of aligned semantic instruments (typically, controlled vocabularies, thesauri, gazetteers, and ontologies) to obtain semantically-"related" terms. For each related term, a new query is generated. The final result consists of a set of semantically-related queries that are all executed by the discovery broker. The query results are clustered according to the semantically-related terms utilised for the discovery expansion. The second discovery style applies the same strategy, but in this case, semantically-related terms, retrieved from the aligned semantic instruments, are presented to users who browse a graph developed according to these terms and select the terms that they deem most pertinent to the search. In both cases, the set of terms that are identified may be further expanded with multiple translations of the terms into other languages.

These modules of the brokering framework have been successfully adopted and implemented in the GEOSS Common Infrastructure (GCI) (Heimbigner and McLeod, 1985), and have resulted in a dramatic increase in the number of resources made available to users (GEOSS, 2012), from a few hundreds to tens of millions. The reason is simple: Brokering represents a shift in philosophy from asking many disciplines and communities to make costly changes and adopt new standards and protocols, to an approach of asking them to make no changes but only declare which standards and protocols they use. The Brokering then takes the responsibility of building the necessary bridges and delivering extended usability to the contributing communities and disciplines.

Links to Societal Benefits

Blue Planet is an important new initiative in the framework of GEOSS. GEOSS in turn is the flagship project of GEO that was launched in response to calls for action by the 2002 World Summit on Sustainable Development and by the G8 (Group of Eight) leading industrialised countries. These high-level meetings recognised that international collaboration is essential for exploiting the growing potential of Earth observations to support decision making in an increasingly complex and environmentally stressed world. GEOSS aims to address nine "Societal Benefit Areas" of disasters, health, energy, climate, water, weather, ecosystems, agriculture and biodiversity. During its first 10 years of implementation (2005-15) GEOSS has mobilised the scientific community in each of these areas, and developed a Common Infrastructure to improve the sharing of data needed at the global level to underpin progress in each of these SBAs. The Brokering framework described in this chapter is an essential component of this common infrastructure as it makes it possible to share these data across disciplines and communities in an unprecedented way. By way of example, David (2012) described how the combined analysis of the ocean physical parameters, with other environmental and socio-economic data, allows determining the best location of intensive fish farming without damaging the environment. Repeating this excellent research at a global scale requires the integration of multiple datasets from different disciplines and organisations, which would not be possible without the brokering framework described in this chapter as different organisations collect and manage their data in different ways and with different standards and protocols.

Conclusion

As noted at the beginning of this chapter, several issues must be addressed to finalise a seamless and comprehensive system of comparable and timely observations of the oceans. While Brokering does not address the quality and comparability of the source data, it helps a great deal in bridging gaps across multi-disciplinary communities and supporting a holistic and integrated assessment of problems and solutions. This beautifully complements and makes effective the on-going process of standards adoption and systems federation, and makes it considerably more effective.

CHAPTER TWENTY-SEVEN

CONCLUSIONS

SAMY DJAVIDNIA, SOPHIE SEEYAVE AND TREVOR PLATT

This book reflects not only the outcome of the "Oceans and Society: Blue Planet" Kick-Off Symposium, which took place in Ilhabela, São Paulo, Brazil from 19–21 November 2012, but also provides readers with additional elements describing the future development of the GEO Task.

Participants at the Symposium agreed on three immediate follow-up actions:

1. Continue developing the Blue Planet Task and establish further synergies between its various Task Components;
2. Prepare a Mission Statement and White Paper to explain the contributions of the various programmes and elements of the Blue Planet Task (see Annex I);
3. Publish the key contributions of the Symposium in the form of a book entitled "Oceans and Society: Blue Planet".

All of these objectives have been met. The Task has been developed through the establishment of a website (http://www.oceansandsociety.org) and a Steering Committee, and through a more rational allocation of programme elements among the Components of the Blue Planet. New elements, such as ocean acidification, have been added, as well as a new Component for the Coastal Zone. The Mission Statement and the White Paper have been produced, as may be seen in this book.

Following the Symposium, the "Oceans and Society: Blue Planet" Task and outcomes of the Symposium were presented by Albert Fischer (IOC, UNESCO) at the GEO-IX Plenary in Foz do Iguaçu, Brazil (22–23 November 2012). The presentation was drafted during the Blue Planet Symposium with input from all the participants, and was very well received by the GEO Plenary with many supporting interventions from various delegations: Australia, Brazil, France, Italy, Madagascar, Norway, United States, European Union, CEOS, COSPAR, International Society for Digital Earth (ISDE), and POGO). The Plenary wholeheartedly supported the Blue Planet initiative, and its relevance to the commitments made at Rio+20 was highlighted.

Next steps in the development of Blue Planet

Knowledge transfer is paramount. Services (information and products) are key to realising the true value of the data, rather than the data themselves. We therefore need to provide appropriately tailored science-based ocean services to individuals and organisations to improve their decision-making and policy processes. There is no value in providing information that the "end users" do not need or do not understand. The needs and requirements of these "end users" cannot, and should not, be taken for granted; we need to offer services that are fit for purpose and fit for use.

We need to engage society with creative, innovative and enterprising programmes that can ignite and inspire the next generation of ocean and social scientists. There must be a coordinated programme for ocean education and outreach aimed at the "citizen", showcasing the wonder of our oceans. In addition, we need to nurture a two-way relationship with society, one where citizens contribute to the science and, for example, form an integral part of the ocean observation community through social media technology. Where relevant, we may also need to adapt our capacity-building programmes to include social scientists, communication specialists, and the "end-users". Capacity building is already strong in Blue Planet, but we should be alert to the possibility of extending the outreach to society at large.

By engaging with society, the ocean community can more accurately determine the new technologies required to respond to policy needs. We must be daring and look 20 years ahead to appreciate what the needs may be and how we can develop science and technology to respond.

Continuous investment in scientific excellence is required. National and international medium-to-long-term plans need to be established to secure the highest level of scientific excellence in ocean sciences. Ultimately we will engage successfully with society only if investments enabling scientists to work on research and development in a sustainable manner are made.

The ultimate goal is to have a "Blue Planet" Task which is, as much as possible, society-oriented. Through the activities of the "Oceans and Society: Blue Planet" initiative, we aim to contribute to the sustainability of our ocean ecosystem. This book, including the "Oceans and Society: Blue Planet" White Paper (see Annex I), provides a starting point for meeting the challenges we face now, and those we expect to face in the future.

ANNEX I

THE BLUE PLANET WHITE PAPER

Background

As established in 2005, the intergovernmental Group on Earth Observations (GEO) is structured around nine societal-benefit areas (SBAs). The ocean was not named as one of the SBAs, although it is easy to show that the oceans play a role, often a fundamental one, in all the nine SBAs identified by GEO: climate, weather, biodiversity, ecosystems, agriculture, health, disasters and water. For example, observations of the ocean are critical for monitoring climate variability and change, and for generating forecasts and projections of climate that can be used in climate services. Ocean observations help improve predictions of longer-range forecasts of weather. Ocean-related hazards such as tsunamis, storm surges, and extreme waves require ocean observations for early warning systems and to prepare for and mitigate the effects of disasters. Because of their role in climate, ocean observations also provide important information for the forecasts of precipitation and drought, the source of replenishment of water supplies, and of climate events that can lead to public health incidents or changes in energy demand. Ocean biological observations are critical in monitoring the health of ocean ecosystems and biodiversity, and the way ecosystem services are being impacted by a changing environment. They are also important in managing fisheries, which fall into the Agriculture SBA. Finally, oceans impinge on various cross-cutting initiatives in GEO, such as system architecture and capacity building.

Although various successful marine activities were conducted under earlier GEO Tasks, they were uncoordinated and often overlooked. In short, oceans did not receive the prominence they deserved. Furthermore, the distribution of the tasks in multiple SBAs did not facilitate integration and synergy between elements. The Partnership for Observation of the Global Oceans (POGO) worked hard to have this situation reversed, so that oceans would be accorded their proper place in the GEO arena. In addition, from a communication point of view, it was felt that speaking with a common voice for the oceans would be more effective than fragmented messages.

In parallel, there was a growing desire to integrate various marine observing programmes in other quarters. For example, in 2010, the European Space Agency (ESA) prepared a report to GEO which stated the case very well:

> "What is needed now, that GEOSS will help achieve, is to integrate the outputs from these various marine monitoring and observation efforts into a cohesive system of systems which will enable researchers, resource managers and policy makers to rapidly assess what is known about a particular marine region…".

Against this background, in 2010, POGO produced the draft prospectus for a new overarching GEO Task to be devoted entirely to oceans. Over the next two years, the new Task was developed further by POGO, in collaboration with other marine interests, notably the Global Ocean Observing System (GOOS), the Committee on Earth Observing Satellites (CEOS), and GODAE OceanView. It was introduced formally into the 2012 Work Plan of GEO (Oceans and Society: Blue Planet). The new Task seeks to coordinate the marine work under GEO and develop synergies between its various components, with a view to the generation of information for societal benefit. A related goal is to raise awareness of the societal importance of observing the ocean, both *in situ* as well by remote sensing.

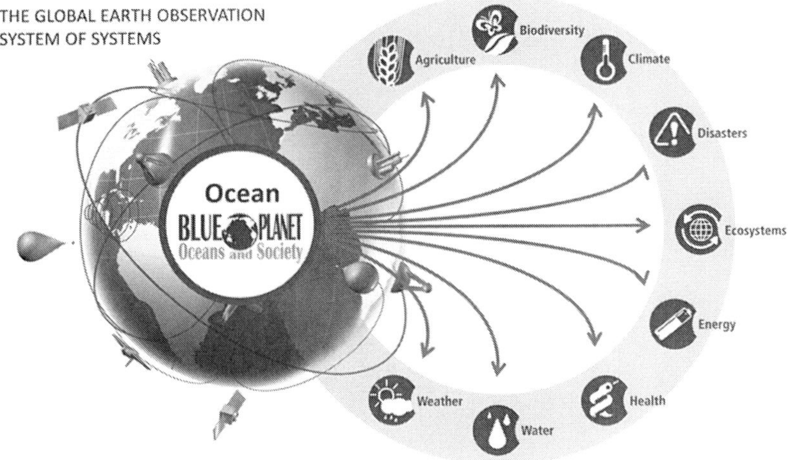

Figure AI-1: Links between the Blue Planet Task and the GEO SBAs

Mission Statement

Oceans and Society: Blue Planet seeks, through the mobilisation of expert knowledge, to raise public awareness of the role of the oceans in the Earth system, of their impacts (good and bad) on humankind, and of the societal benefits of ocean observations; to coordinate the various marine initiatives within GEO and develop synergies between them; and to advocate and advance the establishment and maintenance of a global observing network for the oceans, which acknowledges the value of ocean observations and their contribution to helping alleviate societal issues in multiple areas.

Structure

The Blue Planet Task is divided into six Components. They are:

C1: Sustained Ocean Observations
C2: Sustained Ecosystems and Food Security
C3: Ocean Forecasting and Services
C4: Services for the Coastal Zone
C5: Ocean Climate and Carbon
C6: Developing Capacity and Social Awareness

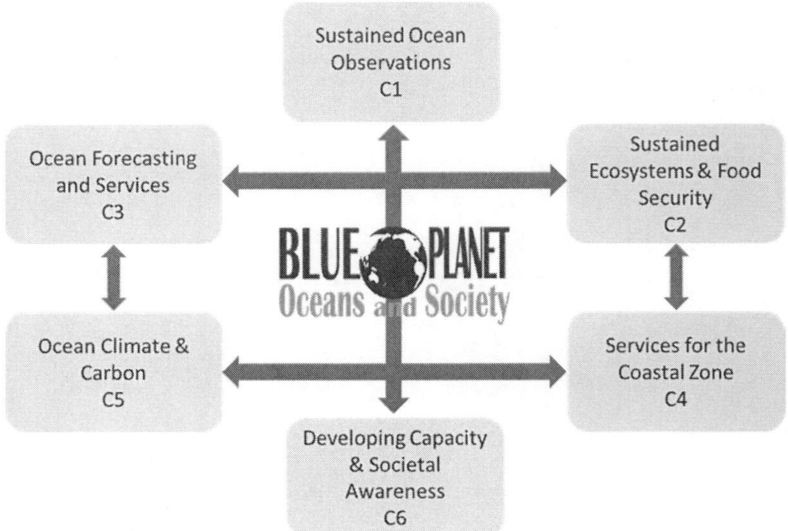

Figure AI-2: Links between the Blue Planet Task six Components

Internally, Blue Planet links strongly together the activities of POGO, GOOS, CEOS, GODAE OceanView and the Coastal Zone Community of Practice (CZCP).

Figure AI-3: Organisation leading the Blue Planet Task Components

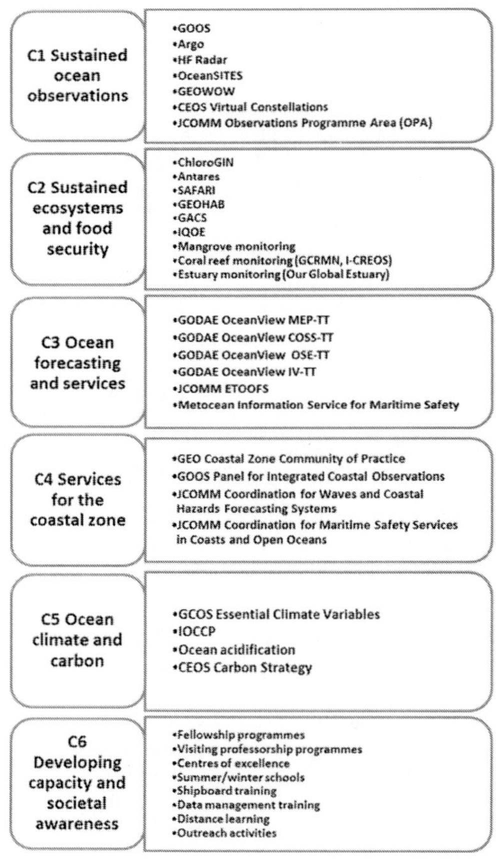

Figure AI-4: Elements contained in the six Blue Planet Task Components

Components of Blue Planet

C1: Sustained Ocean Observations

The overall goal of this Component is to deliver a sustained ocean observing system meeting societal and scientific needs for data and information. More specifically, it will:

- Develop the IOC-WMO-UNEP-ICSU Global Ocean Observing System (GOOS) as a voluntary collaborative system based on the Framework for Ocean Observing and building as much as possible on coordinated efforts using existing structures and in partnership with other organisations such as POGO, GEO, and the Joint WMO-IOC Technical Commission for Oceanography and Marine Meteorology (JCOMM) Observations Programme Area (OPA);

- Sustain and develop present global observations of physical variables for climate, weather, and ocean forecasting based on national contributions adhering to common best practices and standards and an open data policy, and benefiting from common technical coordination and support (through JCOMM; and promoting standards and practices consistent with GEOSS Common Infrastructure, to facilitate access to additional users);
- Promote cooperation of space agencies in measuring Essential Ocean Variables and Essential Climate Variables through the CEOS Virtual Constellations for sea surface temperature, ocean surface topography, surface vector wind, and ocean colour;
- Facilitate coordination of relevant activities of institutions and organisations with *in situ* observing capabilities through POGO;
- Expand GOOS to address new societal challenges through definition of requirements for additional Essential Ocean Variables including biological and biogeochemical variables (note link to C2), coordination of observing networks jointly with POGO and CEOS and appropriate data management arrangements. The WMO Rolling Review of Requirements (RRR) will be a main tool to review continuously and update the observing requirements for open ocean and coasts;
- Develop cooperation between governmental and non-governmental ocean data management initiatives (including the IOC International Oceanographic Data and Information Exchange IODE, the Ocean Biogeographic Information System OBIS, GOOS Regional Alliance data management arrangements, and specific regional arrangements such as GMES/Copernicus), and identify their contribution to the GEOSS Common Infrastructure; and
- Develop the capacity of nations at the global scale to contribute to, and benefit from, ocean observations, with a particular focus on developing services and products driven by local needs and requirements. This has a clear link to C6.

The key outputs from the Component will be:

- Sustained ocean data streams of Essential Ocean Variables from satellite and *in situ* observing networks;

- Revised statements of requirements for ocean Essential Climate Variables and developed statements of requirements for geochemical and biological/ecosystems Essential Ocean Variables (cooperative work of WMO/RRR, GCOS, GOOS, GEO BON, POGO, etc.);
- Phased implementation plans (cooperative work of GCOS, GOOS, JCOMM) that target high-feasibility and high-impact observations and developing integration of measurement of variables on platforms where appropriate;
- Metrics for implementation targets and real-time tracking of the status of *in situ* observing arrays (cooperative work of GOOS, JCOMM); and
- Common technical coordination platform to help operators of *in situ* networks, tracking of network status, health and distribution as well as real-time data distributions, development of deployment opportunities for autonomous platforms (drifters, oats), and other general programme support (JCOMM Observing Programme Support Centre: JCOMMOPS).

The essential activity is coordination on behalf of and with the appropriate observing system stakeholders, working to build a common system. These stakeholders include IOC Member State governmental agencies, CEOS space agencies, leaders and representatives of the ocean research and observing community such as POGO, leaders of individual *in situ* observing networks, GOOS Regional Alliances, and representatives of primary output/user communities such as data management activities, ocean forecasters, and ocean assessments.

The Intergovernmental Oceanographic Commission (IOC) of UNESCO provides overall governance to IODE and to GOOS in collaboration with other sponsors WMO, UNEP and ICSU. The GOOS Steering Committee (GSC), providing technical guidance to GOOS, is composed of ocean observing system experts and links to relevant research and observing communities (such as POGO, SCOR, SCAR, WCRP and IGBP). Oversight on global open-ocean observing network of GOOS, which is also regarded as the ocean component of the WMO Integrated Global Observing System (WIGOS), is provided by JCOMM Observations Coordination Programme Area (JCOMM/OPA), which is under the governance of WMO and IOC. These frameworks define and update requirements, refine or implementation plans; and engage and interact with stakeholders.

GOOS Regional Alliances also exist in many regions to coordinate regionally-agreed activities that contribute to GOOS as well as regional and national goals. A number of them focus on coastal observations and promote the sharing of data regionally. Their activities are supported by secretariats based at IOC/UNESCO and other locations, which also work on: supporting fundraising efforts; supporting the building of institutional capacity related to GOOS; and overall outreach and communication on GOOS and related activities. The communication is largely aimed at Member States (government representatives) and the scientific community participating in and benefiting from GOOS.

C2: Sustained Ecosystems and Food Security

The purpose of this Component is to provide sustained, integrated and globally-complete observations of the ocean ecosystem for use first by the scientific community to assess and anticipate possible changes in its structure and function following environmental perturbations such as climate change or overfishing; and second by the decision makers responsible for ocean stewardship. Here, the programmes ChloroGIN (Chlorophyll Globally-Integrated Network) and GACS (Global Alliance of Continuous Plankton Recorder Surveys) have important roles.

ChloroGIN is a worldwide network of experts in the use of ocean-colour data from remote sensing to recover the large-scale chlorophyll field and in the related bio-optical measurements that aid the interpretation of the remotely-sensed data. The remote-sensing component in ChloroGIN is complemented by time-series observations of ecosystem properties, especially bio-optical measurements. ChloroGIN is conceived of as a network of networks, with various regional nodes. It arose from international training courses conducted in Latin America by POGO, jointly with the IOCCG. The Latin American network (Antares) was extended globally with support from GEO and GOOS. ChloroGIN is recognised as a pilot project of GOOS; hence the link to C1 is obvious. Since many of the time series stations participating in ChloroGIN are coastal stations, this activity is also linked closely to C5.

The globally-scoped GACS has developed from a survey programme that originated in the UK and is now more than eighty years old. These surveys, which are accomplished in partnership with commercial ships, collect detailed information on the plankton community in the ocean, at the species level. Through GACS, the continuous plankton surveys are now being extended to many other oceans.

Another important application of the work in C2 relates to issues in fisheries (including aquaculture) and, ultimately, to security in food from the sea. The topic is covered by the SAFARI (Societal Applications in Fisheries and Aquaculture of Remote-sensing Imagery) Project, which was an original GEO initiative sponsored by the Canadian Space Agency that has proved to be highly successful, with various activities including the first international symposium on fisheries and remote sensing, held in India in 2010. Under SAFARI, the IOCCG has issued a monograph on remote sensing and fisheries, as well as a handbook of case studies for the application of ocean- colour data to problems of ocean management, especially fisheries.

Harmful Algal Blooms (HAB) constitute a problem of increasing importance around the world. Component C2 is linked to GEOHAB (Global Ecology and Oceanography of Harmful Algal Blooms), the principal international programme in this subject area. ChloroGIN is uniquely suited to assist research and operations related to HAB. The C2 Component has clear links with C4 (Services for the Coastal Zone).

Ocean acidification is now considered a serious threat to marine ecosystems. It is an issue that cuts across various Components of Blue Planet.

A major goal of C2 would be to:

- Assure sustainability and global coverage for both ChloroGIN and GACS. The longer-term aim would be to make both fully operational and therefore to transition them from C2 to C1.

C3: Ocean Forecasting and Services

The long-term objective of this Component is to raise capability of ocean forecasting and analysis; in support of the safety of life and property at sea and in coastal areas, risk management for ocean- based economic/commercial/industrial activities, the prevention and control of marine pollution, sustaining healthy and productive oceans and developing integrated coastal area management services. In a wider context, the pursuit of this Component is linked with that of C4.

Scientific and technical coordination for this Component is managed by the GODAE OceanView Science Team (GOVST), whose objectives are to define, monitor, and promote actions aimed at coordinating and integrating research associated with multi-scale and multidisciplinary ocean analysis and forecasting systems, thus enhancing the value of these systems for research and applications. The science team provides international coordination and leadership in:

- Consolidation and improvement of global and regional analysis and forecasting systems (physics);
- Progressive development and scientific testing of the next generation of ocean analysis and forecasting systems, covering bio-geochemical and eco-systems as well as physical oceanography, and extending from the open ocean into the shelf sea and coastal waters;
- Exploitation of this capability in other applications (such as weather forecasting, seasonal and decadal prediction, climate change detection and its coastal impacts); and
- Assessment of the contribution of the various components of the observing system and scientific guidance for improved design and implementation of the ocean observing system.

The GOVST coordinates and implements its work-plan through the nationally and internationally funded activities of its members.

At the national level, the main deliverables are concerned with improvements to the quality and scope of the ocean monitoring and prediction systems; services based on the outputs of these systems; and exploitation of these services. At the international level, the main deliverables are annual reports briefly describing the national systems and services and their exploitation; information concerning the impact of elements of the observing system on the forecast products; workshops facilitating coordination and collaboration; and a quinquennial conference and journal special issue summarising progress.

The GOVST meets about once a year to exchange information about progress at national levels, to coordinate the work of the Task Teams and to plan major events. Five Task Teams have been established to pursue particular activities. A Project Office with one member of staff supports these teams. The five teams are:

- Intercomparison and validation of the forecasts;
- Experiments to evaluate the impact of the observing system on the forecasts;
- Application to the Coastal Oceans and Shelf Seas (COSS);
- Application for Marine Ecosystem Analysis and Prediction (MEAP); and
- Short- to medium-range Coupled Weather Prediction (SRCP).

Building on the success of GODAE and continuous development through GOV, operational ocean forecast capability has now been implemented in a number of countries. Such developments have allowed significant improvement of the oceanographic and met-ocean services, to the point of reliable operational services, such as meteorological services. The Services and Forecasting Systems Programme Area (SFSPA) of JCOMM coordinate global efforts to implement operational ocean forecasting services, through; 1) developing technical guidance for modelling and data assimilation; 2) improved data flow between real time observations, new observations and forecasting and analysis systems; 3) enhanced coordination of review requirements as well as performance verification for operational forecasting systems and services; 4) coordination of the development of coupled ocean-atmosphere forecasting systems and climate coupled models as needed. The operational services are developed and provided by those national and regional operational centres that are capable of operating and maintaining ocean modelling and forecasting systems.

An important activity to this end, carried out by the JCOMM/SFSPA Expert Team on Operational Ocean Forecasting Systems (ETOOFS) is to develop a Guide to Operational Ocean Forecasting, and to conduct routine performance monitoring of the forecasting systems. The ETOOFS and GOVST work closely with each other, to ensure seamless transition of technologies into operation, and to keep abreast of the state-of-art knowledge in planning, producing and assessing the services and products.

One of the most established areas of ocean services is meteorological information for maritime safety; the basic concept of the Global Maritime Distress and Safety System (GMDSS) is that search and rescue authorities ashore, as well as shipping in the immediate vicinity of the ship in distress, will be alerted rapidly so that search and rescue operations can be made with the minimum of delay. Weather and sea information within the GMDSS is provided in the framework of World-Wide Metocean Information and Warning Service (WWMIWS) that is coordinated by the WMO Marine Meteorology and Oceanography Programme (MMOP) with technical advice by JCOMM/SFSPA, in parallel with the World-Wide Navigation Warning Service (WWNWS) coordinated by the International Maritime Organization (IMO) and the International Hydrographic Organization (IHO), within the International Convention for the Safety of Life at Sea (SOLAS). An important application of ocean forecasting and analysis is to provide more reliable and timely information on dangerous sea state in open ocean and coasts (e.g., Ports), through the WWMIWS framework.

C4: Services for the Coastal Zone

The goal of this Component is to improve access to environmental intelligence for all stakeholders in the coastal zone, and to support deliberations on coastal zone management as well as decision making related to the sustainable development of the coastal zone. It will facilitate adaptive, ecosystem-based approaches (EBAs) to sustainable development, including marine spatial planning and management. Sustainable development depends on the continued provision of ecosystem goods and services valued by society. EBAs require the sustained provision of multidisciplinary data (biogeochemical and ecological as well as geophysical) and information on ecosystems states, especially in the coastal zone where goods and services are most in demand.

Coastal zones are areas of particular ecological, social, and economic value where many conflicting interests need to be resolved to ensure sustainable development of this highly complex environment. Humankind has always shown a special interest in the coastal zone and a large fraction of the human population traditionally settles in or close to the coastal zone. Recent increases in coastal population and urbanisation and changes in land-use practices have led to rapid and large changes in sediment supplies and increases in nutrient, pollutant and pathogen loadings to coastal waters. Climate-induced changes in sea level are likely to increase the risk of inundation in many parts of the coastal zone. The on-going and anticipated changes pose serious risks to society and to the capacity of ecosystems to support products and services critical to the survival and well-being of human populations, in developed and developing nations alike.

The Component has the goal to provide observation-based intelligence required for making informed decisions concerning the coastal zone. High and immediate priorities for GEOSS are services that inform about coastal hazards, including but not limited to improved forecasts of local coastal sea- level rise and associated increases in coastal inundation that may be exacerbated by increases in the frequency of extreme weather. Likewise, services focusing on the state and potential degradation of coastal ecosystems, including loss and modification of crucial habitats such as mangroves, coral reefs and estuaries, as well as the state of other environmental conditions, including water quality (e.g., coastal eutrophication and hypoxia; human exposure to waterborne pathogens; harmful algal blooms), are key contributions needed to provide accurate and timely support for decision making for sustainable coasts.

A key deliverable of this Component is the Global Coastal Zone Information System (GCZIS), which will be implemented as a globally-available cyber infrastructure and populated for a few selected study cases. The assessment of observational requirements and demonstrators for a set of monitoring services, including ones for mangroves, coral reefs, estuaries, water quality, sea level, and changes in coastal morphology are also deliverables. Likewise, implementation will be pursued of a regional coastal pilot project in a priority "super site" domain (e.g., Indonesian Archipelago-South China Sea domain), in accordance with a PICO/Coastal GOOS Report (2012) recommendation. This will demonstrate the value added of an end-to-end system of systems for ecosystem-based approaches for monitoring and managing the coastal zone. The end-to-end system of systems pilot project to support ecosystem-based approaches (EBAs) for the coastal zone will be coordinated with GOOS (especially the relevant GOOS Regional Associations) and other global and regional observing system networks and entities as articulated in the PICO Plan.

The GCZIS will be a resource for all other Components of the Blue Planet Task: coordination with, and feedback from, the other Components will be important. Other proposed elements of this Component, including the sea level and water quality monitoring services, as well as the integrated regional coastal pilot project, will benefit from infrastructure provided by Components 1, 2 and 3 (e.g., remote and *in situ* observations as well as well as emerging forecasting capabilities).Ecosystem-related services identified here and in the supporting Task document, e.g., mangrove, coral reef and estuary monitoring, have been suggested for inclusion into Component 2 and will address gaps and provide needed services to coastal users; a strong link between these activities and the new Coastal Services Component will be important.

Recognising the extreme vulnerability of coastal zones to natural disasters particularly inundation caused by multiple and combined phenomena; Disaster Risk Reduction (DRR) is the key priority in developing and application of the GCZIS. The focus is to improve capabilities for operational monitoring and forecasts/warnings on coastal inundation from combined extreme events, such as extreme sea level rise (e.g., large waves, storm surges, high tide), fluvial flooding and tropical cyclones, and furthermore, for decision support system for emergency management. Such efforts require an integrated approach by multidisciplinary communities. To this end JCOMM, jointly with the WMO Technical Commission for Hydrology (CHy), guides the implementation of WMO Coastal Inundation Forecasting Demonstration Project (CIFDP). This project aims to support countries to establish an integrated forecasting system on coastal inundation based on interdisciplinary collaboration between scientists, operational forecasters, coastal disaster managers and institutional end-users. The CIFDP implemented at a national level by operational forecasting agencies, under the WMO framework and with technical guidance provided by JCOMM and CHy. The CIFDP has a clear link to C6.

C5: Ocean Climate and Carbon

The goal of the Ocean Climate and Carbon Component of Blue Planet is to advance the development and implementation of the marine component of the observation systems for both Climate and Carbon, and in particular to address the issues and synergies across the climate-carbon interface for the marine environment. There exist substantive and relevant programmes and coordinating mechanisms addressing climate and carbon observations for the oceans. The added-value of this Blue Planet Component will be to:

- Provide additional exposure and visibility to these programmes, where possible making a concerted effort to link to relevant policies and policy frameworks;
- Identify additional synergies between the programmes; and
- Provide additional linkages with complementary activities within Blue Planet as well as inside and outside of the GEO community.

For both climate and carbon there are concrete requirements on the basis of which the relevant activities will be developed and monitored.

- For the Ocean Climate Observations the requirements are represented by the ocean Essential Climate Variables of the Global Climate Observing System; and
- For Ocean Carbon, the requirements and observing system recommendations are summarised in the GEO Carbon Strategy report, which makes specific recommendations for the ocean domain.

On the explicit basis of these requirements various coordinating bodies and programmes are making substantive efforts: the Blue Planet Task will take advantage of these. For the space agency coordination, CEOS is working in both the Climate and Carbon areas, and is benefiting in both from the cross cutting competences that come from the ocean Virtual Constellations. For climate the ocean ECVs are dealt with through the CEOS Working Group on Climate and can benefit from the structural work undertaken through the activities on the definition of a Climate Monitoring Architecture (joint CEOS-CGMS-WMO activity). For Carbon, the CEOS Carbon Task Force is analysing a response to the GEO Carbon Strategy in which the chapter on oceans evidences critical implementation requirements. The GCOS Implementation Plan also addresses in-situ requirements for the Ocean domain. General oversight of these activities is monitored through the Ocean Observation Panel for Climate and many of the in-situ monitoring networks for ocean cli- mate observations are coordinated by JCOMM. For the carbon ECVs, and the biogeochemical observations in general, the International Ocean Carbon Coordination Project (IOCCP) promotes the development of a global network of ocean carbon observations for research through technical coordination and communication services, international agreements on standards and methods, and advocacy and links to the global observing systems.

Finally there are also relevant data synthesis and consolidation activities such as the SOCAT (Surface Ocean CO2 Atlas) which could make a significant contribution to this Component. There are also specific initiatives addressing ocean acidification, which, in the Blue Planet Task, is dealt with by the Ocean Carbon and Climate Component. In fact, during the past few years, multiple national and international research projects on ocean acidification have emerged, significantly advancing the knowledge base in this domain. To lever the greatest value from these initial research investments and results, the establishment of an international coordination platform on ocean acidification was suggested by the SOLAS-IMBER Ocean Acidification Working Group (SIOA WG) and the International Ocean Acidification Reference User Group (IOA-RUG). An Ocean Acidification International Coordination Centre (OA-ICC) was officially launched by the IAEA in June 2012. The Ocean Climate and Carbon Component will benefit within Blue Planet from the strong emphasis of tasks and components addressing marine ecosystems and biogeochemistry, as well as from the obvious linkages between the climate observations and operational oceanography observations and modelling. Furthermore there will be a fundamental need to maintain links with the domain generic tasks under the Climate SBA of the GEO Work Plan on advancing the Climate Information for Adaptation and that on Global Carbon Observation and Analysis. Initial efforts on this Component of Blue Planet will largely address the interfaces between many of the subject areas and communities described above, including:

- Addressing the interface between the relevant Climate and Carbon communities for ocean applications;

- Addressing the interface between activities coordinating the space-based and in-situ observations for both climate and carbon;
- Ensuring that the sustained observation capacity for both climate and carbon is continuously and appropriately motivated, to the appropriate funding bodies, and that this is done through a consistent framework such as that proposed for the Climate Monitoring Architecture;
- Ensuring that adequate integration with the other relevant Blue Planet Components is developed so that an observational basis is available for periodic multi-pressure assessments, to be performed for the global oceans; and
- Developing an information system framework/portal comprising observational data for ocean climate and carbon from both satellite and in-situ observations.

C6: Developing Capacity and Social Awareness

The main purpose of this Component is to maintain, develop and expand capacity-building in the field of ocean observations. It is initiated by first identifying deficiencies in knowledge and skills, and then determining efficient and effective approaches to address these gaps. The principal outcome will be a larger pool of trained personnel, more representative of the world scale of the demand for such personnel, and more highly trained in the collection and interpretation of ocean data, both *in situ* and remotely sensed. The goal will be addressed by a combination of training courses, visiting-scientist fellowships, visiting-professor fellowships, cruise-participation fellowships and a Centre of Excellence in Ocean Observations (currently located at the Alfred Wegener Institute in Germany). Because the ocean transcends national boundaries, and because it is so vast and challenging to access, oceanographic research requires international collaboration by its very nature. Furthermore, capacity building requires at the very least two-way interaction between developing and developed countries, for example through exchange of personnel, knowledge and expertise between countries. Such exchanges can be, and are, facilitated by international organisations such as POGO, IOC and the Scientific Committee on Oceanic Research (SCOR). Capacity building can be most effective through the participation of developing countries in research projects run in collaboration with developed countries. Ultimately, if capacity building is truly successful, developing nations will become more integrated and active in international networks, such as those named above.

International organisations and programmes such as IOC, SCOR, POGO and others work together to ensure that capacity building is carried out in an integrated manner, avoiding duplication of effort. For example, POGO and SCOR collaborate to deliver a Fellowship Programme and the IOC's International Oceanographic Data and Information Exchange (IODE) provides training in data management at the Centre of Excellence in Observational Oceanography that is run by POGO in partnership with the Nippon Foundation. The three organisations meet regularly to discuss a common strategy, identify needs and plan new activities, such as, for example, the setting up of the Ocean Summer Schools web portal in 2011 (http://www.oceansummerschools.org).

The Group on Earth Observations (GEO) also has a strong mandate for capacity building with particular emphasis on societal benefits. Examples include ChloroGIN and SAFARI that both came into being largely through GEO initiatives. ChloroGIN aims to promote *in situ* measurement of chlorophyll in combination with satellite-derived estimates and associated products to facilitate assessment of marine ecosystems for the benefit of society.

Developing social awareness is an outreach activity. It will be addressed in the first instance by the production of a series of videos aimed at the general public. The series debut will be made at the GEO Ministerial in Geneva, January 2014. A longer-term goal is to find ways to facilitate the transfer of information from the observing systems into the decision making function.

Engaging with Partner Organisations

Externally, Blue Planet is connected with a rich assemblage of agencies and programmes. These include intergovernmental commissions, such as IOC and JCOMM, as well as international coordinating mechanisms and programmes for global systems, such as GOOS and GCOS (co- sponsored by the international science community represented in ICSU), international scientific panels such as GOV, as well as agency-user panels, such as CEOS and the IOCCG. Some of the links are shown in Figure AI-5.

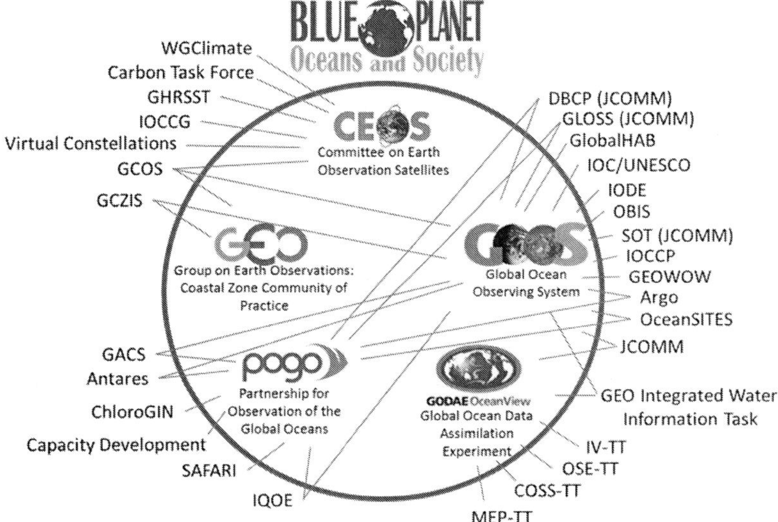

Figure AI-5: Links between the Blue Planet Task Organisations and Programmes

Of particular note is the collaboration of Blue Planet with the European Commission (EC) in its efforts to promote a transatlantic initiative in ocean observing with North America (Canada and the United States). Here, the Blue Planet has been recognised by the EC as a viable platform on which the cooperation could be based. The EC and Blue Planet will appear together in a joint side event at the GEO-X Plenary (Geneva, January 2014).

The actions taken by the EC regarding transatlantic collaboration are embedded in the larger Horizon 2020 Programme, which includes many funding opportunities relevant to the activities of Blue Planet.

Governance

Most of the elements of Blue Planet already have their own governance structures of varying complexities. The Blue Planet Task seeks to aggregate and integrate these elements, ideally building on existing structures and developing synergies between them where appropriate. Blue Planet does not seek to obliterate or replace these structures, but only to find an integration protocol that would make minimal disturbance. A governance structure is indeed required for Blue Planet itself, but it should, as far as possible, have a light touch.

A Steering Committee has been established with representation from the leaders of the six Components, as well as from agencies and institutions with an interest in the success of the Blue Planet Task. The membership is:

Trevor Platt	POGO (Chairman)
Mike Bell	Met Office, United Kingdom
Michael Berger	EC
Douglas Cripe	GEO Secretariat
Paul DiGiacomo	NOAA, United States
Samy Djavidnia	Independent (Editor of the "Oceans and Society: Blue Planet" book)
Mark Dowell	EC-JRC
Albert Fischer	GOOS
Boram Lee	JCOMM
Pierre-Philippe Mathieu	ESA
Shubha Sathyendranath	PML, United Kingdom
Kerry Sawyer	CEOS
Andy Steven	CSIRO, Australia

The Terms of Reference for this committee are to:

1. Oversee and guide the evolution of the Blue Planet Task.
2. Promote Blue Planet in appropriate international scientific fora.
3. Constitute the *pro tem* governance structure for the Blue Planet Task.
4. Promote the societal applications of ocean observations.
5. Facilitate the interconnections between elements of the Blue Planet Task and their linkages with relevant programmes and activities.
6. Aid delivery of benefits to society from Blue Planet.
7. Assist in securing a resource base for execution of the Blue Planet Task.

Funding

Various elements of Blue Planet already have their own funding base. Although these are often considered to be insufficient, it remains true that some funds are available, and in many cases have been available for some time to advance various elements of Blue Planet. That being said, it has to be acknowledged that, at present, there is no central funding for the Blue Planet Task: coordination, development and reporting proceed on a best-effort basis, as is the case for other Tasks in GEO. To a certain extent,

Blue Planet is subsidised discreetly by other programmes. This is not a sustainable situation.

As examples of the levels of funding behind some of the elements of Blue Planet, we might mention GOOS (in Component C1), at roughly $500,000 per annum; and the POGO capacity-building programme (in Component C6) in excess of $600,000 per annum. But these are the exceptions rather than the rule. Many outstanding programmes grouped under Blue Planet have no funds whatsoever. For example, neither ChloroGIN nor SAFARI (in Component C2), both widely-praised, has any direct funding at present. At the international level, funding is also required for the overall coordination of the complex Blue Planet Task.

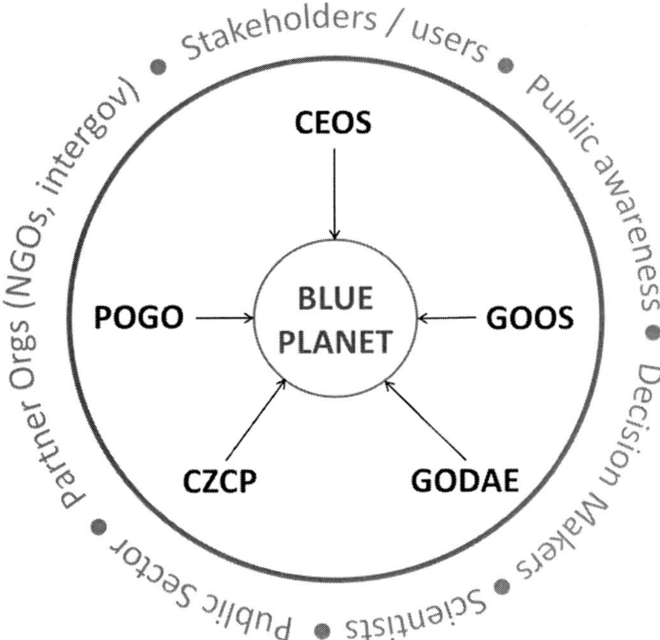

Figure AI-6: The Blue Planet Task supports the provision of ocean services for the benefit of society

Added Value of Blue Planet

The Blue Planet Task would be of limited value if it were not more than just an aggregation of existing elements. There should also be added value. In fact, work under the Task is already developing valuable synergies between the elements. Among the factors that contribute to the added value of Blue Planet are that it:

- Provides new platforms to demonstrate importance of sustained *in situ* and satellite observations of marine and freshwater environments, and the value of integrating these with models;
- Brings together a wide and diverse community of governmental and academic researchers and provides new platforms for integration of multiple streams of data into products that provide the real value to users (such as sustainable fisheries);
- Includes a dynamic, focussed programme in capacity building complemented by a vigorous, global network of former scholars from developing countries;
- Brings the ocean observing community's efforts to the view of GEO decision-makers and funders;
- Provides opportunities to build on GEOSS infrastructure from other domains (ocean data have their greatest value when combined with information from other domains, including socio-economic data);
- Coordinates with other GEO projects that have a marine component, such as GEOBON; and
- Encourages the scientific ocean observing community to demonstrate the value of their observations for the societal benefit areas of GEO.

There is an excitement, momentum and growing interest in the Task, which is helping to carry it forward. The first Blue Planet Symposium, in Ilhabela, Brazil on the 19-21 November 2012, brought together more than 70 ocean community experts representing 24 countries and organisations. Since then, major new elements have requested to be admitted, such as a global network on ocean acidification and the Coastal Zone Community of Practice (which has led to the inclusion of a completely new Component, C4). The Blue Planet Task is clearly answering a real need.

It is also clear that the Blue Planet Task resonates strongly with issues of importance to society, applying the fruits of innovative science and the priceless results of an integrated observing system to provide ocean services for the benefit of society and thus enable sustainable development of the Blue Economy (Figure AI-6).

Two particular opportunities present themselves. The first is to use the POGO-led vehicle Oceans United to accelerate the outreach and communication of the results of ocean observations to decision makers. In this respect, POGO is a neutral, sober and widely-respected "voice of the oceans", unfettered by the constraints of governmental, intergovernmental and diplomatic ways of working. In fact, POGO has pioneered the creation of the Blue Planet on behalf of Oceans United, in other words on behalf of the entire marine community. Many of the elements of Blue Planet, and of the partner organisations, are already members of Oceans United, and those who are not yet members would be encouraged to join. Oceans United is a vehicle that can be developed much further.

The other opportunity is a rapidly-developing initiative between the European Commission and North America (Canada and the United States) for Trans-Atlantic cooperation in observing the ocean. Here, Blue Planet has been recognised, especially by the EC, as a viable platform on which the cooperation could be based. Recognising that many of the issues that confront us today (such as climate change) are global in scope, Blue Planet will strive to ensure that the architecture of the North Atlantic observing system will be such that the results can be integrated seamlessly into those of the world data bases, thus optimising the return on investment.

www.oceansandsociety.org
@OceansandSociety

ANNEX II

ACRONYMS AND GLOSSARY

AC	Atmospheric Correction
ACC	Antarctic Circumpolar Current
ACORN	Australian Coastal Ocean Radar Network
ADCP	Acoustic Doppler Current Profiler
AERONET	AErosol RObotic NETwork
AMESD	African Monitoring of Environment for Sustainable Development
AMOC	Atlantic Meridional Overturning Circulation
AMT	Atlantic Meridional Transect
ANDS	Australian National Data Service
ANN	Artificial Neural Network
AODN	Australian Oceanographic Data Network
AOML	Atlantic Oceanographic and Meteorological Laboratory
ATLAS	Autonomous Temperature Line Acquisition System
AVHRR	Advanced Very High Resolution Radiometer
BGC	BioGeochemical Cycles
BIO	Bedford Institute of Oceanography
BoM	Bureau of Meteorology (Australia)
CA	Coastal Altimetry
CDOM	Coloured Dissolved Organic Matter
CDR	Climate Data Record
CEOS	Committee on Earth Observation Satellites
CGMS	Coordination Group for Meteorological Satellites
ChloroGIN	Chlorophyll Globally Integrated Network
CHy	Technical Commission for Hydrology
CIFDP	Coastal Inundation Forecasting Demonstration Project
CLIVAR	CLImate VARiability and predictability
CNES	Centre National d' Etudes Spatiales (National Centre for Space Studies, France)
CODAR	Coastal ocean dynamics applications radar

COFS	Coastal Ocean Forecasting Systems
CoPs	Communities of Practice
COOP	Coastal Ocean Observing Panel
COSPAR	The Committee on Space Research
COSS	Coastal Ocean and Shelf Seas
COSS-TT	Coastal Ocean and Shelf Seas Task Team
CPR	Continuous Plankton Recorder
CPUE	Catch Per Unit Effort
CSA	Canadian Space Agency
CSIRO	Commonwealth Scientific and Industrial Research Organisation (Australia)
CSSWG	Coastal and Shelf Seas Working Group (GODAE)
CTF	Carbon Task Force (CEOS)
CZCP	Coastal Zone Community of Practice
CZCS	Coastal Zone Colour Scanner
D&I	Data and Information
DA	Data Assimilation
DBCP	Database Connection Pool
DIVERSITAS	An international programme of biodiversity science
DOC	Dissolved Organic Carbon
EAMNet	Europe Africa Marine Earth Observation Network
EBAs	Ecosystem Based Approaches
EC	European Commission
ECV	Essential Climate Variable
EMODNet	European Marine Observation and Data Network
EMSA	European Maritime Safety Agency
EMSO	European Multidisciplinary Seafloor Observatory
ENSO	El Niño Southern Oscillation
ENVISAT	ENVIronmental SATellite
EO	Earth Observation
ESA	European Space Agency
ETOOFS	Expert Team for Operational Ocean Forecast Systems
EUMETSAT	European Organisation for the Exploitation of Meteorological Satellites
FLH	Fluorescence Line Height
FOO	Framework for Ocean Observing
FR	Frequently Repeated
GA	Geosciences Australia
GACS	Global Alliance of Continuous Plankton Recorder
GARP	Global Atmosphere Research Programme

GBR	Great Barrier Reef
GCOS	Global Climate Observing System
GCRMIN	Global Coral Reef Monitoring Network
GCZIS	Global Coastal Zone Information System
GGOS	Global Geodetic Observing System
GTOS	Global Terrestrial Observing System
GEO	Group on Earth Observations
GEO BON	Group on Earth Observations Biodiversity Observation Network
GEOHAB	Global Ecology and Oceanography of Harmful Algal Blooms
GEOSS	Global Earth Observation System of Systems
GEOSS Data-CORE	GEOSS Data Collection of Open Resources for Everyone
GEOWOW	GEOSS interoperability for Weather, Ocean and Water
GGOS	Global Geodetic Observing System
GHRSST	Group on High Resolution Sea Surface Temperature
GIA	Glacial Isostatic Adjustment
GISS	Goddard Institute for Space Studies (NASA)
GLEON	Global Lake Ecological Observatory Network
GLOSS	Global Sea Level Observing SyStem
GMDSS	Global Maritime Distress and Safety System
GMES	Global Monitoring for Environment and Security
GOCI	Geostationary Ocean Colour Imager
GODAE	Global Ocean Data Assimilation Experiment
GOOS	Global Ocean Observing System
GOV	GODAE Ocean View
GOVST	GODAE Ocean View Science Team
GPS-RTK	Global Positional Service – Real Time Kinematic
GRA	GOOS Regional Alliance
GRACE	Gravity Recovery and Climate Experiment
GSOP	Global Synthesis and Observations Panel
GTOS	Global Terrestrial Observing System
GTS	Global Telecommunication System
HAB	Harmful Algal Bloom
HICO	Hyperspectral Imager for the Coastal Ocean
HD	High Density
HF	High Frequency
HFSWR	High-Frequency Surface Wave Radar

IAEA	International Atomic Energy Agency
IAI	Inter-American-Institute for global change research
ICAN	International Coastal Atlas Network
ICES	International Council for the Exploration of the Sea
I-CREOS	International Network of Coral Reef Ecosystem Observing Systems
ICSU	International Council for Science (previously International Council of Scientific Unions)
ICT	Information and Communications Technologies
IGARSS	International Geoscience And Remote Sensing Symposium
IGBP	International Geosphere-Biosphere Programme
IGCO	Integrated Global Carbon Observing
IGST	International GODAE Science Team
IGOS	Integrated Global Observing Strategy
IIOE	International Indian Ocean Experiment
IMBER	Integrated Marine Biogeochemistry and Ecosystem Research
IMOS	Integrated Marine Observing System (Australia)
INCOIS	Indian National Centre for Ocean Information Services
INSITU-OCR	International Network for Sensor InTer-comparison and Uncertainty assessment for Ocean Colour Radiometry
IOA-RUG	International Ocean Acidification Reference User Group
IOC	Intergovernmental Oceanographic Commission (UNESCO)
IOCCG	International Ocean-Colour Coordinating Group
IOCCP	International Ocean Carbon Coordination Project
IODE	International Oceanographic Data and information Exchange
IOOS®	Integrated Ocean Observing System
IOP	Inherent Optical Property
IO-USP	Institute of Oceanography of the University of Sao Paulo (Brazil)
IPCC	Intergovernmental Panel on Climate Change
IQOE	International Quiet Oceans Experiment
ISDE	International Society for Digital Earth
IV-TT	Intercomparison and Validation Task Team

JCOMM	Joint Technical Commission for Oceanography and Marine Meteorology (WMO-IOC)
JCOMMOPS	JCOMM *in situ* Observing Platform Support Centre
JRC	Joint Research Centre (European Commission)
LD	Low Density
LJCO	Lucinda Jetty Coastal Observatory (Australia)
LMR	Living Marine Resource
MEAP	Marine Ecosystem Analysis and Prediction
MEP-TT	Marine Ecosystem Analysis and Prediction
MERIS	MEdium Resolution Imaging Spectrometer
MESA	Monitoring of Environment and Security in Africa
MMOP	Marine Meteorology and Oceanography Programme
MOC	Meridional Overturning Circulation
MODIS	ModeratE Resolution Imaging Spectroradiometer (United States)
MPA	Marine Protected Area
NAO	North Atlantic Oscillation
NANO	NF-POGO Alumni Network for Oceans
NASA	National Aeronautics and Space Administration (United States)
NCRIS	National Collaborative Research Infrastructure Strategy (Australia)
NetLake	Networking Lake Observatories in Europe
NF	Nippon Foundation
NIO	National Institute of Oceanography (India)
NIOZ	Royal Netherlands Institute for Sea Research
NOAA	National Oceanic and Atmospheric Administration (United States)
NSIDC	National Snow and Ice Data Centre (United States)
OA-ICC	Ocean Acidification International Coordination Centre
OBIS	Ocean Biogeographic Information System
OC	Ocean Colour
OCM	Ocean Colour Monitor
OCR-VC	Ocean Colour Radiometry-Virtual Constellation
ODIN	Ocean Data and Information Network
ODIP	Ocean Data Interoperability Platform
OLCI	Ocean Land Colour Instrument
ONR	Office of Naval Research (United States)
OOI	Ocean Observatories Initiative

OOPC	Ocean Observation Panel for Climate
OSE	Observing System Experiment
OSE-TT	Observing System Evaluation Task Team
OSSE	Observing System Simulation Experiment
OST	Ocean Surface Topography
OSVW	Ocean Surface Vector Wind
OLCI	Ocean and Land Colour Instrument
PACE	Pre-Aerosol, Clouds, and ocean Ecosystem
PAR	Photosynthetically Active Radiation
PICO	Panel for Integrated Coastal Observations
PIRATA	Prediction and Research Moored Array in the Tropical Atlantic
PMEL	Pacific Marine Environmental Laboratory
PML	Plymouth Marine Laboratory
POGO	Partnership for Observation of the Global Oceans
PFZ	Potential Fishing Zones
PP	Primary productivity
R&D	Research and Development
ROOS	Regional Ocean Observing System
SAFARI	Societal Applications in Fisheries and Aquaculture of Remote-sensing Imagery
SAHFOS	Sir Alister Hardy Foundation for Ocean Science
SAMOC	South Atlantic Meridional Overturning Circulation Program
SBA	Societal Benefit Area
SBIB	Societal Benefits Implementation Board
SCAR	Scientific Committee on Antarctic Research
SCOR	Scientific Committee on Oceanic Research
SeaDAS	SeaWiFS Data Analysis System
SeaWIFS	Sea-viewing WIde Field-of-view Sensor
SGLI	Second Generation Global Imager (Japan)
SLRCoP	Sea Level Rise Community of Practice
SOCAT	Surface Ocean CO_2 Atlas
SOLAS	Surface Ocean Lower Atmosphere Study
SOOP	Ship of Opportunity
SOT	Ship Observations Team
SRCP	Short-Range Coupled Prediction
SSH	Sea Surface Height
SST	Sea Surface Temperature
TEMA	Training, Education and Mutual Assistance (programme)

TERN	Terrestrial Ecosystem Research Network (Australia)
TSG	ThermoSalinoGraph
TZCF	Transition Zone Chlorophyll Front
UNCLOS	United Nations Convention on the Law of the Sea
UNEP	United Nations Environment Programme
UNESCO	United Nations Educational, Scientific and Cultural Organization
UNFCCC	United Nations Framework Convention on Climate Change
URR	User Requirements Registry
VC	Virtual Constellations
VIIRS	Visible Infrared Imager Radiometer Suite
VOI	Value of Information
WCRP	World Climate Research Project
WERA	Wave Radar
WOCE	World Ocean Circulation Experiment
WMO	World Meteorological Organisation
WWMIWS	World-Wide Metocean Information and Warning Service
WWNWS	World-Wide Navigation Warning Service
XBT	eXpendable BathyThermograph

ANNEX III

REFERENCES AND BIBLIOGRAPHY

ACIL Tasman (2010). Report on Economic Value of Earth Observation from Space – A Review of the Value to Australia of Earth Observation from Space. ACIL Tasman.

Alexandridis, T.K., Topaloglou, C.A., Lazaridou, E. and Zalidis, G.C. (2008). The performance of satellite images in mapping aquacultures. Ocean Coast. Manage. 51: 638-644.

Alvera-Azcárate, A., Barth, A. and Weisberg, R.H. (2009). A nested model of the Cariaco Basin (Venezuela): description of the basin's interior hydrography and interactions with the open ocean. Ocean Dynamics 59(1), 97-120.

Amorin, F.N., Cirano, M., Soares, I.D., Campos, E.J.D. and Middleton, J.F. (2012). The influence of large-scale circulation, transient and local processes on the seasonal circulation of the Eastern Brazilian Shelf, 13 S. Cont. Shelf Res. 32: 47-61.

Anderson, C.R., Kudela, R.M., Benitez-Nelson, C., Sekula-Wood, E., Burrell, C.T., Chao, Y., Langlois, G., Goodman, J. and Siegel, D.A. (2011). Detecting toxic diatom blooms from ocean color and aregional ocean model, Geophys. Res. Lett. 38(4), DOI: 10.1029/2010GL045 858.

Anderson, C.R., Seigel, D.A., Kudela, R. and Brzezinski, M.A. (2009). Empirical models of toxigenic *Pseudo-nitzschia* blooms: Potential use as a remote detection tool in the Santa Barbara Channel, Harmful Algae 8: 478–492.

Anderson, D. M., Cembella, A.D. and Hallegraeff, G. M. (2012a). Progress in understanding Harmful Algal Blooms: Paradigm shifts and new technologies for research, monitoring, and management. Ann. Rev. Mar. Sci. 4: 143–176.

Anderson, D.M., Doucette, G., Scholin, C.A., Paul, J., Trainer, V.L., Campbell, L., Kudela, R.M., Stumpf, R.P. and Morrison, J.R. (2012b). Harmful algal bloom (HAB) sensors in ocean observing systems, Interagency Ocean Observation Committee summit white paper.

Andrews, G.E. (2012). Drowning in the data deluge. Not. Am. Math. Soc. 59: 913-941.

Antoine, D., Morel, A., Gordon, H. R., Banzon, V. F. and Evans, R.H. (2005). Bridging ocean color observations of the 1980s and 2000s in search of long-term trends. J. Geophys. Res. 110, C06009, doi:10.1029/2004JC002620.

Atech (2000). Cost of algal blooms. Report to Land and Water Resources Research and Development Corporation, Canberra, ACT 2601 ISBN 0 642 76014 4.

Aufdenkampe, A.K., Mayorga, E., Raymond, P.A., Melack, J.M., Doney, S.C., Alin, S.R., Aalto, R.E. and Yoo, K. (2010). Riverine coupling of biogeochemical cycles between land, oceans, and atmosphere. Front. Ecol. Environ. 9: 53–60, 10.1890/100014.

Australian Government (2013). Australia's Satellite Utilisation Policy. Australian Government – Department of Industry, ISBN 978-1-922218-16-2, DIISRTE 12/257.

AVISO (2013). Mean Sea Level issues: questions in discussion. http://www.aviso.oceanobs.com/en/news/ocean-indicators/mean-sea-level/msl-science-issues.html#c9012.

Backer, L.C. and McGillicuddy, D.J. (2006). Harmful algal blooms. At the interface between coastal oceanography and human health. Oceanography 19: 94–106.

Baird, M.E., Ralph, P.J., Rizwi, F., Wild-Allen, K. and Steven, A.D.L. (2013). A dynamic model of the cellular carbon to chlorophyll ratio applied to a batch culture and a continental shelf ecosystem. Limnol. Oceanogr. 58(4): 1215-1226.

Baker, J.D., Polovina, J.J. and Howell, E.A. (2007). Effect of variable oceanic productivity on the survival of an upper trophic predator, the Hawaiian monk seal Monachus schauinslandi. Mar. Ecol. Prog. Ser. 346: 277-283.

Bakun, A. and Broad, K. (2003). Environmental 'loopholes' and fish population dynamics: comparative pattern recognition with focus on El Niño effects in the Pacific. Fish. Oceanogr. 12: 458–473.

Balmaseda, M.A., Trenberth K.E. and Källen, E.E. (2013). Distinctive climate signals in re-analysis of global ocean heat content. Geophys. Res. Lett. 40: 1754-1759.

Bastviken, D., Cole, J.J., Pace, M.L. and Tranvik, L. (2004). Methane emissions from lakes: Dependence of lake characteristics, two regional assessments, and a global estimate. Glob. Biogeochem. Cycles 18, 10.1029/2004GB002238.

Battin, T.J., Kaplan, L.A., Findlay, S., Hopkinson, C.S., Marti, E., Packman, A.I., Newbold, J.D. and Sabater, F. (2008). Nature Geoscience 1: 95-100.

Battin, T.J., Luyssaert, S., Kaplan, L.A., Aufdenkampe, A.K., Richter, A. and Tranvik, L.J. (2009) The boundless carbon cycle. Nature Geoscience 2: 598-600.

Bell, M.J., Lefèbvre, M., Le Traon, P.-Y., Smith, N. and Wilmer-Becker, K. (2009). GODAE: The Global Ocean Data Assimilation Experiment. Oceanography 22(3): 14–21, doi:10.5670/oceanog.2009.62.

Bell, M.J., Lefèbvre, M., Le Traon, P-Y., Smith, N. and Wilmer-Becker, K. (2009). GODAE: The Global Ocean Data Assimilation Experiment. Oceanography 22(3), http://www.tos.org/oceanography/archive/22-3_le_traon.html.

Bigagli, L., Nativi, S., Mazzetti, P. and Villoresi, G. (2004). GI-Cat: a Web Service for Dataset Cataloguing Based on ISO 19115. In: Proceedings of the 15th International Workshop on Database and Expert abase and Expert Systems Systems Applications, Zaragoza (Spain), Sep. 2004, IEEE Computer Society Press, ISBN 0-7695-2195-9, pp. 846-850.

Bigagli, L., Papeschi, F., Nativi, S., Bastin, L. and Masó, J. (2013). Ensuring consistency and persistence to the Quality Information Model-The role of the GeoViQua Broker, EGU General Assembly Conference Abstracts, Vol. 15, pp. 8789.

Bindoff, N., Willebrand J., Artale V., Cazenave A., Gregory J., Gulev S., Hanawa K., Le Quéré, C., Levitus, S., Nojiri, Y., Shum, C.K., Talley, L.D. and Unnikrishnan, A. (2007). Observations: Oceanic climate change and sea level. In: Climate Change 2007: The Physical Science Basis. Contribution of Working Group I to the Fourth Assessment Report of the Intergovernmental Panel on Climate Change, Solomon, S., Qin, D., Manning, M., Chen, Z., Marquis, M., Averyt, K.B., Tignor, M. and Miller, H.L. (eds.), Cambridge University Press, Cambridge, pp. 385-342.

Bleck, R. (2006). On the use of hybrid vertical coordinates in ocean circulation modeling. In: Ocean Weather Forecasting. An Integrated View of Oceanography, Chassignet, E.P. and Verron, J. (eds.), Springer, pp. 109-126.

Blondeau-Patissier, D., Brando, V.E., Oubelkheir, K., Dekker, A.G., Clementson, L.A. and Daniel, P. (2009). Bio-optical variability of the absorption and scattering properties of the Queensland inshore and reef waters, Australia. J. Geophys. Res. 114, 10.1029/2008JC005039.

Boersma, M., Malzahn, A.M., Greve, W. and Javidpour, J. (2007). The first occurrence of the ctenophore *Mnemiopsis leidyi* in the North Sea. Helgol. Mar. Res., 10.1007/s10152-006-0055-2.

Bograd, S., Foley, D.G., Schwing, F.B., Wilson, C., Polovina, J.J. and Howell, E.A. (2004). On the seasonal and interannual migrations of the Transition Zone Chlorophyll Front. Geophys. Res. Lett., 10.1029/2004GL020637.

Borgman, C.L., Wallis, J.C. and Enyedy, N. (2007). Little science confronts the data deluge: habitat ecology, embedded sensor networks, and digital libraries. Int. J. Dig. Libr. 7: 17-30.

Borstad, G., Crawford, W., Hipfner, J.M., Thomson, R. and Hyatt, K. (2011). Environmental control of the breeding success of rhinoceros auklets at Triangle Island, British Columbia. Mar. Ecol. Prog. Ser. 424: 285-302.

Bouma, J.A., Kouk, O. and Dekker, A.G. (2011). Assessing the value of Earth Observation for managing coral reefs: An example from the Great Barrier Reef. Sci. Total Environ. 409: 4497–4503.

Bourlès, B., Lumpkin, R., McPhaden, M.J., Hernandez, F., Nobre, P., Campos, E., Yu, L., Planton, S., Busalacchi, A., Moura, A.D., Servain, J. and Trotte, J. (2008). The PIRATA program: History, accomplishments, and future directions. Bull. Amer.Meteor. Soc. 89: 1111–1125.

Boutin, J., Etcheto, J., Dandonneau, Y., Bakker, D.C.E., Feely, R.A., Inoue, H.Y., Ishii, M., Ling, R.D., Nightingale, P.D., Metzl, N. and Wanninkhof, R. (1999). Satellite sea surface temperature: a powerful tool for interpreting *in situ* pCO_2 measurements in the equatorial Pacific Ocean. Tellus B 51(2): 490–508, 10.1034/j.1600-0889.1999.00025.x.

Brando, V.E., Anstee, J.M., Wettle, M., Dekker, A.G., Phinn, S.R. and Roelfsema, C. (2009). A physics based retrieval and quality assessment of bathymetry from suboptimal hyperspectral data. Remote Sens. Environ., 10.1016/j.rse.2008.12.003.

Brando, V.E., Schroeder, T., Dekker, A.G. and Clementson, L. (2013). Reef Rescue Marine Monitoring Program: Using remote sensing for GBR-wide water quality. Final Report for 2011/2012 Activities. Canberra, ACT, CSIRO Land and Water.

Brando, V.E., Dekker, A.G., Park, Y.J., and Schroeder, T. (2012). An adaptive semianalytical inversion of ocean colour radiometry in optically complex waters. Applied Optics 51(15): 2808-2833.

Brando, V.E., Keen, R., Daniel, P., Baumeister, A., Nethery, M., Baumeister, H., Hawdon, A., Swan, G., Mitchell, R., Campbell, S., Schroeder, T., Park, Y.J., Edwards, R., Steven, A., Allen, S., Clementson, L. and Dekker, A. (2010). The Lucinda Jetty Coastal Observatory's role in Satellite Ocean colour calibration and validation for Great Barrier Reef coastal waters. IEEE Oceans'10 Sydney, 25-27 May 2010.

Brewin, R.J.W., Hardman-Mountford, N.J., Lavender, S.J., Raitsos, D.E., Hirata, T., Uitz, J., Devred, E., Bricaud, A., Ciotti, A. and Gentili, B. (2011). An intercomparison of bio-optical techniques for detecting phytoplankton size class from satellite remote sensing. Remote Sens. Environ. 115(2): 325-339.

Cazenave, A. and Llovel, W. (2010). Contemporary Sea Level Rise. Annu. Rev. Mar. Sci., 2, 145-173, 10.1146/annurev-marine-120308-081105.

Cazenave, A. and Nerem, R.S. (2004). Present-Day sea level change: observations and causes. Rev. Geophys. 42, RG3001, 8755-1209/04/2003RG000139.

Cazenave, A. and Remy, F. (2011). Sea level and climate: measurements and causes of changes. WIREs Clim. Change 2: 647-662, 10.1002/wcc.139.

Cazenave, A., Henry, O., Munier, S., Delcroix, T., Gordon, A.L., Meyssignac, B., Llovel, W., Palanisamy, H. and Becker, M. (2012). Estimating ENSO influence on the global mean sea level over 1993-2010. Mar. Geod. 35(sup1): 82-97, 10.1080/01490419.2012.718209.

CEOS (2014). The CEOS Strategy for Carbon Observations from Space, Draft version 1.2 available from http://ceos.org/index.php?option=com _content&view=category&layout=blog&id=159&Itemid=204.

Chassignet, E.P., Hurlburt, H.E., Metzger, E.J., Smedstad, O.M., Cummings, J.A., Halliwell, G.R., Bleck, R., Baraille, R., Wallcraft, A.J., Lozano, C., Tolman, H.L., Srinivasan, A., Hankin, S., Cornillon, P., Weisberg, R., Barth, A., He, R., Werner, F. and Wilkin, J. (2009). US GODAE: Global ocean prediction with the HYbrid Coordinate Ocean Model (HYCOM). Oceanography 22(2): 64–75, http://dx.doi. org/10.5670/oceanog.2009.39.

Chen, I.-C., Lee, P.-F. and Tzeng, W.-N. (2005). Distribution of albacore (*Thunnus alalunga*) in the Indian Ocean and its relation to environmental factors. Fish. Oceanogr. 14: 71-80.

Christian, R.R., P.M. DiGiacomo, T.C. Malone, and L. Talaue-McManus. (2006). Opportunities and challenges of establishing coastal observing systems. Estuaries and Coasts 29: 871–875.

Church, J.A., White, N.J., Konikow, L.F., Domingues, C.M., Cogley, J.G., Rignot, E., Gregory, J.M., van den Broeke, M.R., Monaghan, A.J. and Velicogna, I. (2011). Revisiting the Earth's sea-level and energy budgets from 1961 to 2008. Geophys. Res. Lett. 38, L18601, 10.1029/ 2011GL048794.

Church, M.J., Lomas, M.W. and Muller-Karger, F. (2013). Sea change: charting the course for biogeochemical ocean time-series research in a new millennium. Deep-Sea Res.II, 93: 2-15.

Ciais, P., Dolman, A.J., Dargaville, R., Barrie, L., Butler, J., Canadell, J. and Moriyama, T. (2010). GEO Carbon Strategy. Rome: GEO Secretariat, Geneva / FAO.

Ciborra, C.U. and Patriota, G. (1998). Groupware and teamwork in R&D: limits to learning and innovation. Res. Dev. Manage. 28(1): 1-10.

Cipollini, P., Benveniste, J., Miller, L., Picot, N., Scharroo, R., Strub, T., Vandemark, D., Vignudelli, S., Zoffoli, S., Andersen, O., Bao, L., Birol, F., Coelho, E., Deng, X., Emery, W., Fenoglio, L., Fernandes, J., Gomez-Enri, J., Griffin, D., Han, G., Hausman, J., Ichikawa, K., Kostianoy, A., Kourafalou, V., Labroue, S., Ray, R., Saraceno, M., Smith, W., Thibaut, P., Wilkin, J. and Yenamandra, S. (2012). Conquering the coastal zone: a new frontier for satellite altimetry. Proceedings of the "20 Years of Progress in Radar Altimetry" Symposium, Venice, Italy, Benveniste, J. and Morrow, R. (eds.), 9/2012, ESA Special Publication SP-710 (CD-ROM), ESA Publications Division, European Space Agency, Noordwijk, The Netherlands.

Clark, H.L. and Isern, A. (2003). Oceanography 16(4):5, http://dx.doi.org/ 10.5670/oceanog.2003.01.

Cole, J.J., Prairie, Y.T., Caraco, N.F., McDowell, W.H., Tranvik, L.J., Striegl, R.G., Duarte, C.M., Kortelainen, P., Downing, J.A., Middelburg, J.J. and Melack J. (2007). Plumbing the Global Carbon Cycle: Integrating Inland Waters into the Terrestrial Carbon budget. Ecosystems 10: 171-184.

Comiso, J.C. (2012). Large decadal decline of the Arctic multiyear ice cover. J. Climate 25: 1176-1193, 10.1175/JCLI-D-11-00113.1.

Cooper, M. and Haines, K. (1996). Altimetric assimilation with water property conservation. J. Geophys. Res. 101(C1): 1059-1077.

Copertino, M.S., Garcia A.E., Muelbert, J.H. and Garcia, C.A. (2010). Introduction to the Special Volume on Climate Changes and Brazilian Coastal Zones. Pan-American Journal of Aquatic Sciences 52(3): i-vi.

Counillon, F. and Bertino, L. (2009). High-resolution ensemble forecasting for the Gulf of Mexico eddies and fronts. Ocean Dynamics 59(1): 83-95.

Cox, P.M., Betts, R.A., Jones, C.D., Spall, S.A. and Totterdell, I.J. (2000). Acceleration of global warming due to carbon-cycle feedbacks in a coupled climate model. Nature 408: 184–187.

CSIRO (2012). Continuity of Earth Observation Data for Australia: Research and Development Dependencies to 2020. Commonwealth of Australia.

Cushing, D.H. (1969). The regularity of the spawning season of some fishes. J. Phys. Oceanogr. 33: 81-92.

Cushing, D.H. (1975). Marine Ecology and Fisheries. Cambridge University Press, Cambridge, UK, 278 pp.

Cushing, D.H. (1990). Plankton production and year-class strength in fish populations – an update of the match mismatch hypothesis. Adv. Mar. Biol. 26: 249-294.

David, L. (2012). Endeavouring for Wise Mariculture: Science Input to Policy Implementation. Presentation at the Blue Planet Kick-Off Meeting, Ilhabela, Brazil, 19 November 2012, http://www.oceans andsociety.org/files/david.pdf.

De Mey, P. and Proctor, R. (2009). Assessing the value of GODAE products in coastal and shelf seas. Editorial, Special Issue of Ocean Dynamics, 2007 GODAE Coastal and Shelf Seas Workshop, Liverpool, UK. Ocean Dynamics 59: 1–2, 10.1007/s10236-008-0175-0.

De Mey, P., Craig, P., Davidson, F., Edwards, C.A., Ishikawa, Y., Kindle, J.C., Proctor, R., Thompson, K.R., Zhu, J. and the GODAE Coastal and Shelf Seas Working Group (CSSWG) community (2009). Applications in coastal modelling and forecasting. Oceanography 22(3):198-205.

De Mey, P., Craig, P., Kindle, J., Ishikawa, Y., Proctor, R., Thompson, K. and Zhu, J. (2007). Towards the assessment and demonstration of the value of GODAE results for coastal and shelf models and forecasting systems. Global Ocean data Assimilation Experiment (GODAE) Coastal and Shelf Seas Working Group (White paper, pp. 79), http:// godae.org/modules/documents/documents/GODAE-CSSWG-paper-ed2.pdf.

Dekker, A.G. and Hestir, E.L. (2012). Evaluating the Feasibility of Systematic Inland Water Quality Monitoring with Satellite Remote Sensing. CSIRO, Australia.

https://publications.csiro.au/rpr/download?pid=csiro:EP117441&dsid= DS10.

DeLong, D. and Fehey, L. (2000). Diagnosing cultural barriers to knowledge management. Acad. Manage. Exec. 14(4): 113-127.

Denman, K.L., Brasseur, G., Chidthaisong, A., Ciais, P., Cox, P.M., Dickinson, R.E., Hauglustaine, D., Heinze, C., Holland, E., Jacob, D., Lohmann, U., Ramachandran, S., da Silva Dias, P.L., Wofsy, S.C. and Zhang, X. (2007). Couplings Between Changes in the Climate System and Biogeochemistry. In: Climate Change 2007: The Physical Science Basis. Contribution of Working Group I to the Fourth Assessment Report of the Intergovernmental Panel on Climate Change, Solomon, S., Qin, D., Manning, M., Chen, Z., Marquis, M., Averyt, K.B., Tignor, M. and Miller, H.L. (eds.), Cambridge University Press, Cambridge, United Kingdom and New York, NY, USA.

Derber, J. and Rosati, A. (1989). A Global Oceanic Data Assimilation System, J. Phys. Oceanogr. 19(9): 1333-1347.

Devred, E., Sathyendranath, S. and Platt, T. (2009). Decadal changes in ecological provinces of the Northwest Atlantic Ocean revealed by satellite observations. Geophys. Res. Lett. 36, L19607, 10.1029/ 2009GL039896.

Dodds, W.K., Bouska, W.W., Eitzmann, J.L., Pilger, T.J., Pitts, K.L., Riley, A.J., Schloesser, J.T. and Thornbrugh, D.J. (2009). Eutrophication of U.S. freshwaters: analysis of potential economic damages. Environ. Sci. Technol. 43: 12–19.

Domenico, B., Caron, J., Davis, E., Nativi, S. and Bigagli, L. (2006). GALEON: standards-based web services for interoperability among Earth sciences data systems, Geoscience and Remote Sensing Symposium, 2006. IGARSS 2006. IEEE International Conference, pp. 313-316.

Dong, S.S., Garzoli, L., Baringer, M.O., Meinen, C.S. and Goni, G.J. (2009). The Atlantic Meridional Overturning Circulation and its northward heat transport in the South Atlantic. Geophys. Res. Lett. 36, L20606, doi:10.1029/2009GL039356.

Downing, J.A., Prairie, Y.T., Cole, J.J., Duarte, C.M., Tranvik, L.J., Striegl, R.G., McDowell, W.H., Kortelainen, P., Caraco, N.F., Melack, J.M. and Middelburg, J. (2006). The global abundance and size distribution of lakes, ponds, and impoundments. Limnol. Oceanogr. 51: 2388-2397.

Earp, A., Hanson, C.E., Ralph, P.J., Brando, V.E., Allen, S., Baird, M., Clementson, L., Daniel, P., Dekker, A.G., Fearns, P.R.C.S., Parslow,

J., Strutton, P.G., Thompson, P.A., Underwood, M., Weeks, S., and Doblin, M.A. (2011). Review of fluorescent standards for calibration of *in situ* fluorometers: Recommendations applied in coastal and ocean observing programs. Opt. Express, 19(27): 26768-26782.

EarthCube Brokering Concept Award Team (2012). Brokering for EarthCube Communities: A Road Map [Online]. http://api.ning.com/files/e1f7JW76fC4atlM7lc-wdrsh4bpyUb9knJhLXfEtBqjhhF*rgAdvZ NTwzvtnWsqVSXA4EBbE6pNrBO*vpLKR1EiwePStjpnL/Brokering RoadmapAug171.pdf.

Edler, L., Fernö, S., Lind, M.G., Lundberg, R. and Nilsson, P.O. (1985). Mortality of dogs associated with a bloom of the cyanobacterium *Nodularia spumigena* in the Baltic Sea. Ophelia 24: 103–109.

Enkel, E. (2002). Knowledge Networks – an integrated approach to managing explicit and implicit knowledge. The Third European Conference on Organizational Knowledge, Learning and Capabilities (p. 23). Athens: http://apollon1.alba.edu.gr/OKLC2002/Proceedings/pdf_files/ID360.pdf.

European Commission (2012). Green Paper – Marine Knowledge 2020 – from seabed mapping to ocean forecasting. Luxembourg: Publications Office of the European Union, 23 pp., ISBN 978-92-79-25350-8, doi:10.2771/4154.

Evensen, G. (2009). The Ensemble Kalman Filter for Combined State and Parameter Estimation Monte Carlo techniques for data assimilation in large systems. IEEE Control Syst. Mag. 29(3): 83-104.

FAO Fisheries Department (2004). The State of World Fisheries and Aquaculture (SOFIA), U.N. Food and Agriculture Organization, Rome, Italy.

Farias, E.G.G., Nobre, P., Lorenzzetti, J.A., de Almeida, R.A.F. and Júnior, L.C.I. (2013). Variability of air-sea CO_2 fluxes and dissolved inorganic carbon distribution in the Atlantic Basin: a coupled model analysis. Int. J. Geosci. 4: 249-258.

Fetterer, F., Knowles, K., Meier, W. and Savoie, M. (2002, updated 2009). Sea Ice Index. Boulder, CO: National Snow and Ice Data Center. http://dx.doi.org/10.7265/N5QJ7F7W.

Fischlin, A., Midgley, G.F., Price, J.T., Leemans, R., Gopal, B., Turley, C., Rounsevell, M.D.A., Dube, O.P., Tarazona, J. and Velichko, A.A. (2007). Ecosystems, their properties, goods, and services. Climate Change 2007: Impacts, Adaptation and Vulnerability. Contribution of Working Group II to the Fourth Assessment Report of the Intergovernmental Panel on Climate Change, Parry, M.L., Canziani,

O.F., Palutikof, J.P., van der Linden, P.J. and Hanson, C.E. (Eds.), Cambridge University Press, Cambridge, pp. 211-272.

Fiedler, P.C. and Bernard, H.J. (1987). Tuna aggregation and feeding near fronts observed in satellite imagery. Cont. Shelf Res. 7: 871-881.

Firth, L.B., Mieszkowska, N., Thompson, R.C. and Hawkins, S.J. (2012). Climate change and adaptational impacts in coastal systems: the case of sea defences. Environ. Sci. Proc. Imp., DOI: 10.1039/c1033em 00313b.

Fritz, S., Scholes, R.J., Obersteiner, M., Bouma, J. and Reyers, B. (2008). A conceptual framework for assessing the benefits of a Global Earth Observation System of Systems, IEEE Systems Journal 2(3): 338- 348.

Frolov, S., Kudela, R.M., and Bellingham, J.G. (2012). Monitoring of harmful algal blooms in the era of diminishing resources: A case study of the US west coast. Harmful Algae 21-22: 1-12.

Frost and Sullivan (2010). Eutrophication Research Impact Assessment. WRC Report No. TT 461/10. Water Research Commission, Pretoria, South Africa.

Garcia-Soto C., Pingree, R.D. and Valdés, L. (2002). Navidad development in the southern Bay of Biscay: Climate change and swoddy structure from remote sensing and *in situ* measurements. J. Geophys. Res 107 (C8), doi:10.1029/2001JC001012.

Garcia-Soto C., Vázquez-Cuervo, J., Clemente-Colón, P. and Hernández, F. (2012). Satellite Oceanography and Climate Change. Deep-Sea Res. Pt II 77: 1-9, 10.1016/j.dsr2.2012.07.004.

Garcia-Soto, C. and Pingree, R.D. (2012). Atlantic Multidecadal Oscillation (AMO) and Sea Surface Temperature in the Bay of Biscay and adjacent regions. J. Mar. Biol. Assoc. U.K. 92: 213-234, doi:10.1017/S0025315410002134.

Garzoli, S. L., Baringer, M. O., Dong, S., Perez, R.C., Yao, Q. (2013). South Atlantic meridional fluxes. Deep Sea Res.I 71:21-32.

Garzoli, S.L., Boebel, O., Brydene, H., Fine, R.A., Fukasawa, M., Gladyshev, S., Johnson, G., Johnson, M., MacDonald, A., Meinen, C.S., Mercier, H., Orsi, A., Piola, A., Rintoul, S., Speich, S., Visbeck, M. and Wannikhof, R. (2009). Progressing towards global sustained deep ocean observations. Ocean Obs Book of Abstacts, hdl:10013/epic35776.

GCOS (2010). Implementation Plan for the Global Observing System for Climate in Support of the UNFCCC (No. GCOS-92, WMO/TD-No 1219). Retrieved from http://www.wmo.int/pages/prog/gcos/Publications /gcos-138.pdf.

GEOHAB (2011). GEOHAB Modelling: A Workshop Report, McGillicuddy, D.J. Jr., Glibert, P.M., Berdalet, E., Edwards, C., Franks, P. and Ross, O. (eds). IOC and SCOR, Paris and Newark, Delaware, 85 pp.

Geoscience Australia (2011). Continuity of Earth Observation Data for Australia: Operational Requirements to 2015 for Lands, Coasts and Oceans. Commonwealth of Australia.

GEOSS Implementation 2012 Highlights, Document 6, GEO-IX Plenary. Nov. 22–23, 2012 [Online]. http://www.earthobservations.org/documents/geo_ix/20111122_geoss_implementation_highlights.pdf.

Gitelson, A.A., Gao, B.-C., Li, R.R., Berdnikov, S.V. and Saprygin, V. (2011). Estimation of chlorophyll-a concentration in productive turbid waters using Hyperspectral Imager for the Coastal Ocean – The Azov Sea case study. Environ. Res. Lett. 6, 024023.

Glavovic, B. C. (2013). Coastal Innovation Imperative, *Sustainability*, 5, 934-954; doi:10.3390/su5030934.

Glibert, P.M., Seitzinger, S., Heil, C.A., Burkholder, J.M., Parrow, M.W., Codispoti, L.A. and Kelly, V. (2005). The role of eutrophication in the global proliferation of harmful algal blooms. New perspectives and new approaches. Oceanography 18: 198–209.

Godø, O.R., Samuelsen, A., Macaulay, G.J., Patel, R., Hjollo, S.S., Horne, J., Kaartvedt, S. and Johannessen, J.A. (2012). Mesoscale eddies are oases for higher trophic marine life. Plos One 7(1), doi:10.1371/journal.pone.0030161.

Goni, G.J., Roemmich, D., Molinari, R., Meyers, G., Sun, C., Boyer, T., Baringer M., Gouretski, V., DiNezio, P., Reseghetti, F.,Vissa, G., Swart, S., Keeley, R., Garzoli, S., Rossby, T., Maes, C. and Reverdin, G. (2010). The Ship Of Opportunity Program. In: Proceedings of the "OceanObs'09: Sustained Ocean Observations and Information for Society" Conference (Vol. 2), Venice, Italy, 21-25 September 2009, Hall, J., Harrison D.E. and Stammer, D. (eds.), ESA Publication WPP-306.

Goyens, C., Jamet, C. and Schroeder, T. (2013). Evaluation of four atmospheric correction algorithms for MODIS-Aqua images over contrasted coastal waters. Remote Sens. Environ. 131(0): 63-75.

Gregg, W.W., Casey, N.W., O'Reilly, J.E. and Esaias, W.E. (2009). An empirical approach to ocean color data: Reducing bias and the need for post-launch radiometric re-calibration. Remote Sens. Environ. 113(8): 1598–1612, doi:10.1016/j.rse.2009.03.005.

Hallegraeff, G.M. (2003). Harmful algal blooms: A global review. In: Manual on Harmful Marine Microalgae, Hallegraeff, G.M., Anderson, D.M. and Cembella, A.D. (eds), Paris, UNESCO, pp. 1–22.

Hallegraeff, G.M. (2010). Ocean climate change, phytoplankton community responses, and Harmful Algal Blooms: A formidable predictive challenge. J. Phycol. 46: 220–235.

Halliwell, G.R. Jr., Weisberg, R.H., Hogan, P., Smedstad, O.M. and Cummings, J. (2009). Impact of GODAE products on nested HYCOM simulations of the West Florida Shelf. Ocean Dynamics 59(1): 139-155.

Hansen, J., Ruedy, R., Sato, M. and Lo, K. (2010). Global surface temperature change. Rev. Geophys. 48, RG4004, 10.1029/2010RG000 345.

Hardy A.C. (1956). The Open Sea – Its Natural History (Part I): The World of Plankton. The New Naturalist #34. A Survey of British Natural History, Collins, London.

Hardy A.C. (1959). The Open Sea – Its Natural History (Part II) Fish & Fisheries. New Naturalist #37, Collins, London.

Hawkins, S.J., Birth, L.B., McHugh, M., Poloczanska, E.S., Herbert, R.J.H., Burrows, M.T., Burrows, K.M.A., Moore, P.J., Thompson, R.C., Jenkins, S.R., Sims, D.W., Genner, M.J. and Mieszkowska, N. (2013). Data rescue and re-use: Recycling old information to address new policy concerns. Marine Policy 42: 91-98.

Heimbigner, D. and McLeod, D. (1985). A federated architecture for information management. ACM Trans. Office Inform. Syst. 3(3): 253–278.

Heisler, J., Glibert, P., Burkholder, J., Anderson, D., Cochlan, W., Dennison, W. Dortch, Q., Gobler, C.J., Heil, C., Humphries, E., Lewitus, A., Magnien, R., Marshall, H., Sellner, K., Stockwell, D., Stoecker, D. and Suddleson, M. (2008). Eutrophication and Harmful Algal Blooms: A scientific consensus. Harmful Algae 8: 3-13.

Hempel, G. (1998). Dr N K Panikkar: learning research by doing. Mar. Pol. 22(3): 247–254.

Henson, S.A., Dunne, J.P. and Waniek, J.J. (2009). Decadal variability in North Atlantic phytoplankton blooms. J. Geophys. Res. 114, doi:10.1029/2008JC005139.

Herzfeld, M. (2009). Improving stability of regional numerical ocean models. Ocean Dynamics 59(1): 21-46.

Herzfeld, M., Brinkman, R., Gillibrand, P., Andrewartha, J., Rizwi, F., Tonin, H., Jones, E., Hemer, M., Joehnk, K., Margvelashvili, N., Wild-Allen, K., Robson, B., Mongin, M., Skerratt, J., Baird, M., Furnas, M.

and McKinnon, D. (2012). eReefs WP3 Marine Modelling: Milestone 1 Report. Technical report, CSIRO Marine and Atmospheric Research.

Hill, K.L., Rintoul, S.R., Ridgway, K.R., and Oke, P.R. (2011). Decadal changes in the South Pacific western boundary current system revealed in observations and ocean state estimates, J. Geophys. Res. 116, C01009, doi:10.1029/2009JC005926.

Hoagland, P. and Scatasta, S. (2006). The economic effects of harmful algal blooms. In: Ecology of Harmful Algae, Graneli, E. and Turner, J.T. (eds.). Ecological Studies 189: 391-402.

Hoegh-Guldberg, O., Mumby, P.J., Hooten, A.J., Steneck, R.S., Greenfield, P., Gomez, E., Harvell, C.D., Sale, P.F., Edwards, A.J., Caldeira, K., Knowlton, N., Eakin, C.M., Iglesias-Prieto, R., Muthiga, N., Bradbury, R.H., Dubi, A. and Hatziolos, M.E. (2007). Coral reefs under rapid climate change and ocean acidification. Science 318: 1737-1742.

Holthouse, D. (1998). Knowledge management research issues. Calif. Manage. Rev. 40(3): 277-280.

Horner, R.A., Garrison, D.L. and Plumley, F.G. (1997) Harmful algal blooms and red tide problems on the US west coast. Limnol. Oceanogr. 42: 1076–1088.

Howell, E.A., Kobayashi, D.R., Parker, D.M., Balazs, G.H. and Polovina, J. (2008). TurtleWatch: A data product to aid in the bycatch reduction of loggerhead turtle (*Caretta caretta*) in the Hawaii-based pelagic longline fishery. Endang. Species Res. 5: 267-278.

Hurlburt, H.E. (1984). The potential for ocean prediction and the role of altimeter data. Mar. Geod. 8: 17-66.

IOCCG (2008). Why ocean colour? The societal benefits of ocean-colour technology. In: Platt, T., Hoepffner, N., Stuart, V. and Brown, C. (eds.), Reports of the International Ocean-Colour Coordinating Group, No. 7, IOCCG, Dartmouth, Canada.

IPCC (2007). Impacts, Adaptation and Vulnerability. Contribution of the Working Group II to the Fourth Assessment Report of the Intergovernmental Panel on Climate Change, Parry, M. L., Canziani, O.F., Palutikof, J.P., van der Linden, P.J. and Hanson C.E. (eds), Cambridge University Press, Cambridge, United Kingdom and New York, NY, USA. http://www.ipcc.ch/publications_and_data/ar4/wg2/en/contents.html.

IPCC (2013). Climate Change 2013: The Physical Science Basis. Contribution of Working Group I to the Fifth Assessment Report of the Intergovernmental Panel on Climate Change. Stocker, T.F., Qin D., Plattner G.-K., Tignor M., Allen S.K., Boschung J., Nauels A., Xia Y.,

Bex V. and Midgley P.M. (eds.). Cambridge University Press, Cambridge, United Kingdom and New York, NY, USA, 1535 pp.

Jamet, C., Moulin, C. and Lefevre, N. (2007). Estimation of the oceanic pCO_2 in the North Atlantic from VOS lines in-situ measurements: parameters needed to generate seasonally mean maps. Annales Geophysicae 25(11): 2247–2257.

Jessup, D.A., Miller, M.A., Ryan, J.P., Nevins, H.M., Kerkering, H.A., Mekebri, A., Crane, D.B., Johnson, T.A. and Kudela, R.M. (2009). Mass stranding of marine birds caused by a surfactant-producing red tide. PLoS One 4(2), e4550.

Jochens, A.E., Malone, T.C., Stumpf, R.P., Hickey, B.M., Carter, M., Morrison, R., Dyble, J., Jones, B. and Trainer, V.L. (2010). Integrated ocean observing system in support of forecasting harmful algal blooms. Mar. Technol. Soc. J. 44(6): 99-121.

Jones, E., Parslow, J. and Dowd, M. (2013). Summary of Discussions: Marine Biogeochemical Data Assimilation Symposium, Technical report, CSIRO Marine and Atmospheric Research.

Jones, E.M., Oke, P.R., Rizwi, F. and Murray, L.M. (2012). Assimilation of glider and mooring data into a coastal ocean model. Ocean Modelling 47: 1-13.

Joughin, I., Abdalati, W. and Fahnestock, M. (2004). Large fluctuations in speed on Greenland's Jakobshavn Isbræ glacier. Nature 432: 608-610, 10.1038/nature03130.

Kahru, M., Savchuk, O.P. and Elmgren, R. (2007). Satellite measurements of cyanobacterial bloom frequency in the Baltic Sea: interannual and spatial variability. Mar. Ecol. Prog. Ser. 343: 15–23.

Kallio, K. (1999). Absorption properties of dissolved organic matter in Finnish lakes. Proc. Estonian Acad. Sci. Biol, Ecol. 48: 75-83.

Kallio, K., Kutser, T., Hannonen, T., Koponen, S., Pullianen, J., Vepsäläinen, J. and Pyhälahti, T. (2001). Retrieval of water quality from airborne imaging spectrometry of various lake types in different seasons. Sci. Total Environ. 268: 59-77.

Kallio, K., Attila, J., Härmä, P., Koponen, S., Pulliainen, J., Hyytiäinen, U.-M. and Pyhälahti, T. (2008). Landsat ETM+ images in the estimation of seasonal lake water quality in boreal river basins. Environ. Manage. 42: 511-522.

Karl, D.M. (2010). Oceanic ecosystem time-series programs: Ten lessons learned. Oceanography 23(3):104–125, doi:10.5670/oceanog.2010.27.

Kendall, A.W. Jr. and Duker, G.J. (1998). The development of recruitment fisheries oceanography in the United States. Fish. Oceanogr. 7: 69-88.

Kim, H.G. (2006). Mitigation and controls of HABs. Ecological Studies 189: 327–338.

Kleypas, J.A., Buddemeire, R.W. and Gattuso, J.-P. (2001). The future of coral reefs in an age of global change. Int. J. Earth Sci. 90: 426-437.

Kobayashi, D.R., Polovina, J.J., Parker, D.M., Kamezaki, N., Cheng, I., Uchida, I., Duttton, P. and Balazs, G. (2008). Pelagic habitat characterization of loggerhead sea turtles, *Caretta caretta*, in the North Pacific Ocean (1997-2006): Insights from satellite tag tracking and remotely sensed data. J. Exp. Mar. Biol. Ecol. 356: 96-114.

Koivusalo, M., Pukkala, E., Vartiainen, T., Jaakkola, J.J.K. and Hakulinen, T. (1997). Drinking water chlorination and cancer – A historical cohort study in Finland. Cancer Causes Control 8: 192–200.

Koeller, P., Friedland, K., Fuentes-Yaco, C., Han, G., Kulka, D., O'Reilly, J., Platt, T., Richards, A. and Taylor, M. (2009a). Remote sensing applications in stock assessments. In: Remote Sensing in Fisheries and Aquaculture, Forget, M.H., Stuart, V. and Platt, T. (eds), IOCCG, Dartmouth, Canada, pp. 29-42.

Koeller, P., Fuentes-Yaco, C., Platt, T., Sathyendranath, S., Richards, A., Ouellet, P., Orr, D., Skúladóttir, U., Wieland, K., Savard, L. and Aschan, M. (2009b). Basin-scale coherence in phenology of shrimps and phytoplankton in the North Atlantic Ocean. Science 324: 791–793.

Kourafalou, V.H., Peng, G., Kang, H., Hogan, P.J., Smedstad, O.M. and Weisberg, R.H. (2009). Evaluation of Global Ocean Data Assimilation Experiment products on South Florida nested simulations with the Hybrid Coordinate Ocean Model. Ocean Dynamics 59(1): 47-66.

Kowalewski, M. and Ostrowski, M. (2005). Coastal up- and downwelling in the southern Baltic. Oceanologia 47(4): 453–475.

Kozlov, I.E., Kudryavtsev, V.N., Johannessen, J.A., Chapron, B., Dailidienė, I. and Myasoedov, A.G. (2012). ASAR imaging for coastal upwelling in the Baltic Sea. Advances in Space Research 50(8): 1125–1137, doi:10.1016/j.asr.2011.08.017.

Kraberg, A. and Wiltshire, K.H. (2013). Regime shifts in the marine environment: How do they affect ecosystem services? In: The Mediterranean Sea: Its history and present challenges. Springer Verlag, p 790.

Kraus, S.D., Brown, M.W., Caswell, H., Clark, C.W., Fujiwara, M., Hamilton, P.K., Kenney, R.D., Knowlton, A.R., Landry, S., Mayo, C.A., McLellan, W.A., Moore, M.J., Nowacek, D.P., Pabst, D.A.,

Read, A.J. and Rollan, R.M. (2005). North Atlantic right whales in crisis. Science 309: 561-562.

Kudela, R.M., Horner-Devine, A.R., Banas, N.S., Hickey, B.M., Peterson, T.D., McCabe, R.M., Lessard, E.J., Frame, E., Bruland, K.W., Jay, D.A., Peterson, J.O., Peterson, W.T., Kosro, P.M., Palacios, S.L., Lohan, M.C. and Dever, E.P. (2010). Multiple trophic levels fueled by recirculation in the Columbia River plume. Geophys. Res. Lett., doi:10.1029/2010GL044342.

Kudryavtsev, V., Myasoedov, A., Chapron, B., Johannessen, J.A. and Collard, F. (2012). Imaging mesoscale upper ocean dynamics using synthetic aperture radar and optical data. Journal of Geophysical Research-Oceans 117, doi:10.1029/2011JC007492.

Kutser, T. (2004). Quantitative detection of chlorophyll in cyanobacterial blooms by satellite remote sensing. Limnol. Oceanogr. 49: 2179-2189.

Kutser, T., Pierson, D., Tranvik, L., Reinart, A., Sobek, S. and Kallio, K. (2005a). Using satellite remote sensing to estimate the colored dissolved organic matter absorption coefficient in lakes. Ecosystems 8(6): 709–720, doi:10.1007/s10021-003-0148-6.

Kutser, T., Pierson, D.C., Kallio, K.Y., Reinart, A. and Sobek, S. (2005b). Mapping lake CDOM by satellite remote sensing. Remote Sens. Environ. 94(4): 535–540, doi:10.1016/j.rse.2004.11.009.

Kutser, T., Tranvik, L. and Pierson, D.C. (2009). Variations in colored dissolved organic matter between boreal lakes studied by satellite remote sensing. J. Appl. Remote Sens., 10.1117/1.3184437.

Kutser, T. (2012). The possibility of using the Landsat image archive for monitoring long time trends in coloured dissolved organic matter concentration in lake waters. Remote Sens. Environ. 123: 334-338.

Kwok, R. and Comiso, J.C. (2002). Southern Ocean climate and sea ice anomalies associated with the Southern Oscillation. J. Climate 15: 487-501, 10.1175/1520-0442(2002)015<0487: SOCASI>2.0.CO;2.

Kwok, R. and Rothrock, D.A. (2009). Decline in Arctic sea ice thickness from submarine and ICESat records: 1958-2008. Geophys. Res. Lett. 36, L15501, 10.1029/ 2009GL039035.

Landsberg, J. (2002). Effects of algal blooms on aquatic organisms. Rev. Fish. Sci. 10: 113–190.

Lane, J.Q., Raimondi, P.T. and Kudela, R.M. (2009). Development of a logistic regression model for the prediction of toxigenic *Pseudo-nitzschia* blooms in Monterey Bay, California. Mar. Ecol. Prog. Ser. 383: 37-51.

Laughton, A.S. (2004). Recollections of the International Indian Ocean Expedition. Ocean Challenge 13(1): 18-24.

Laurs, R.M., Fiedler, P.C. and Montgomery, D.R. (1984). Albacore tuna catch distributions relative to environmental features observed from satellites. Deep-Sea Res. Pt A. 31: 1085-1099.

Le Hénaff, M., De Mey, P. and Marsaleix, P. (2009). Assessment of observational networks with the Representer Matrix Spectra method – application to a 3D coastal model of the Bay of Biscay. Ocean Dynamics 59(1): 3-20.

Le, C., Zha, Y., Li, Y., Sun, D., Lu, H. and Yin, B. (2010). Eutrophication of lake waters in China: cost, causes, and control. Environ. Manage. 45(4): 662–668.

Lea, D. (2012). Observation Impact Statements for operational ocean forecasting, Met Office Forecasting R&D Technical Report no. 568, http://www.metoffice.gov.uk/media/pdf/o/6/FRTR568.pdf.

Leadbetter, A.M., Lowry, R.K. and Clements, D.O. (2013). Putting meaning into NETMAR – the open service network for marine environmental data. Int. J. Digital Earth, DOI:10.1080/17538947.2013.781243.

Lefevre, N. and Taylor, A. (2002). Estimating pCO_2 from sea surface temperatures in the Atlantic gyres. Deep-Sea Res. Pt I 49(3): 539–554, doi:10.1016/S0967-0637(01)00064-4.

Le Traon, P-Y. Larnicol, G., Guinehut, S., Pouliquen, S., Bentamy, A., Roemmich, D., Donlon, C., Roquet, H., Jacobs, G., Griffin, D., Bonjean, F., Hoepffner, N. and Breivik, L.-A. (2009). Data assembly and processing for operational oceanography: 10 years of achievements. Oceanography 22(3): 56–69.

Lima, J.A.M., Martins, R.P., Tanajura, C.A.S., Paiva, A.M., Campos, E.J.D., Soares, I.D., Cirano, M., França, G.B. and Obino, R.S. (2013). Design and implementation of the Oceanographic Modeling and Observation Network (REMO) for physical oceanography and ocean forecasting. Rev. Bras. Geofis. In press.

Lindsay, R.W. and Zhang, J. (2005). The thinning of Arctic sea ice, 1988-2003: Have we passed a tipping point? J. Climate 18: 4879-4894, 10.1175/JCLI3587.1.

Liu, D., Keesing, J. K., Xing, Q., and Shi, P. (2009). World's largest macroalgal bloom caused by expansion of seaweed aquaculture in China. Mar. Poll. Bull. 58(6): 888-895.

Lohmann, G. and Wiltshire, K.H. (2012). Winter atmospheric circulation signature for the timing of the spring bloom of diatoms in the North Sea. Mar Biol 159: 2573-2581.

Lombard, A., Cazenave, A., Le Traon, P.Y. and Ishii, M. (2005). Contribution of thermal expansion to present-day sea-level change revisited, Glob. Planet. Change 47: 1-16, 10.1016/jgloplacha.2004.11.016.

Lombard, A., Garric, G. and Penduff, T. (2009). Regional patterns of observed sea level change: insights from a 1/4° global ocean/sea-ice hindcast. Ocean Dyn. 59: 433-449, doi:10.1007/s10236-009-0161-.

Longhurst, A. (1998). Ecological Geography of the Sea. Academic Press, San Diego, London, Boston, New York, Sydney, Tokyo, Toronto, 398 pp.

Longhurst, A., Sathyendranath, S., Platt, T. and Caverhill, C. (1995). An estimate of global primary production in the ocean from satellite radiometer data. J. Plank. Res. 17: 1245-1271.

Lorenzoni, L. and Benway, H.M. (eds) (2013). Report of Global intercomparability in a changing ocean: An international time-series methods workshop, November 28-30, 2012, Ocean Carbon and Biogeochemistry (OCB) Program and International Ocean Carbon Coordination Project (IOCCP), 59 pp.

Lutz, V.A., Barlow, R., Tilstone, G., Sathyendranath, S., Platt, T., Hardman-Mountford, N. (2007). Report of the *in situ* component of the 'Plymouth Chlorophyll Meeting and Workshops (Extended Antares Network)'. Sponsored by GOOS, GEO, PML and POGO. 18-22 September 2006. (Web reference: http://www.antares.ws/files/intercal-report.pdf).

Lynch, T.P., Morello, E.B., Evans, K., Richardson, A.J., Steinberg, C.R., Roughan, M., Thompson, P., Middleton, J.F., Feng, M., Sherrington, R.B., Brando, V.E., Tilbrook, B., Ridgway, K., Allen, S., Doherty, P., Hill, K. and Moltmann, T.C. (subm). IMOS National Reference Stations: a continental scaled physical, chemical and biological coastal observing system. Plos ONE: (submitted).

Macauley, M. (2006). The value of information: Measuring the contribution of space-derived Earth science data to resource management, Space Policy 22(4): 274-282.

Magnus, P., Jaakola, J.J.K, Skrondal, A., Alexander, J., Becher, G., Krog, T. and Dybing, E. (1999). Water chlorination and birth defects. Epidemiology: 513-520.

Mahadevan A., D'Asaro, E., Lee, C. and Perry, M.J. (2012). Eddy-driven stratification initiates North Atlantic spring phytoplankton blooms. Science 337, 54. DOI: 10.1126/science.1218740.

Maina, J., Venus, V., McClanahan, M.R. and Ateweberhan, M. (2008). Modelling susceptibility of coral reefs to environmental stress using remote sensing data and GIS models. Ecol. Modell. 212: 180-199.

Margvelashvili, N., Andrewartha, J., Herzfeld, M., Robson, B.J. and Brando, V.E. (2013). Satellite data assimilation and estimation of a 3D coastal sediment transport model using error-subspace emulators. Environ. Modell. Softw. 40: 191-201.

Marta-Almeida, M., Pereira, J. and Cirano, M. (2011). Development of a pilot Brazilian regional operational ocean forecast system, REMO-OOF. J. Oper. Oceanogr. 4: 3-15.

Matear, R. and Jones, E. (2011). Marine biogeochemical modelling and data assimilation. In: Operational Oceanography in the 21st Century, Matear, R., Jones, E. and Schiller, A. (eds.), Springer, pp. 295-317.

Mattern, J.P., Fennel, K. and Dowd, M. (2012). Estimating time-dependent parameters for a biological ocean model using an emulator approach. J. Mar. Syst. 96-97: 32-47.

Maynard, J.A., Turner, P.J., Anthony, K.R.N., Baird, A.H., Berkelmans, R., Eakin, C.M., Johnson, J., Marshall, P.A., Packer, G.R., Rea, A. and Willis, B.L. (2008). Reef Temp: An interactive monitoring system for coral bleaching using high-resolution SST and improved stress predictors. Geophys. Res. Lett. 35: L05603, doi:05610.01029/02007 GL032175.

McClanahan, T.R., Ateweberhan, M., Sebastia, C.R., Graham, N.A.J., Wilson, S.K., Bruggemann, J.H. and Guillaume, M.M.M. (2007). Predictability of coral bleaching from synoptic satellite and *in situ* temperature observations. Coral Reefs 26: 695-701.

McDonald, T.A. and Komulainen, H. (2005). Carcinogenicity of the chlorination disinfection by-product MX. J. Environ. Sci. Health C. Environ. Carcinog. Ecotoxicol. Rev. 23(2), 163-214.

McGillicuddy, D.J., Townsend, D.W., He, R., Keafer, B.A., Kleindinst, J.L., Li, Y., Manning, J.P., Mountain, D.G., Thomas, M.A. and Anderson, D.M. (2011). Suppression of the 2010 *Alexandrium fundyense* bloom by changes in physical, biological, and chemical properties of the Gulf of Maine. Limnol. Oceanogr. 56(6): 2411–2426.

Meier, M.F., Dyurgerov, M.B., Rick, U.K., O'Neel, S., Pfeffer, W.T., Anderson, R.S., Anderson, S.P. and Glazovsky, A.F. (2007). Glaciers dominate eustatic sea-level rise in the 21st century. Science 317: 1064-1067, 10.1126/science.1143906.

Meinen, C.S., Johns, W.E., Garzoli, S.L., van Sebille, E., Rayner, D., Kanzow, T., and Baringer, M.E. (2013). Variability of the Deep

Western Boundary Current at 26.5°N during 2004-2009. Deep-Sea Res. Pt II 85: 154-168.

Merrifield, M.A. and Maltrud, M.E. (2011). Regional sea level trends due to a Pacific trade wind intensification. Geophys. Res. Lett. 38, L21605, 10.1029/2011GL049576.

Meyssignac B., Salas y Melia, D., Becker, M., Llovel, W. and Cazenave, A. (2012). Tropical Pacific spatial trend patterns in observed sea level: internal variability and/or anthropogenic signature? Clim. Past 8: 787-802, 10.5194/cp-8-787-2012.

Mitrovica, J.X., Gomez, N., and Clark, P.U. (2009). The sea-level fingerprint of West Antarctic collapse. Science 323: 753, 10.1126/science.1166510.

Mumby, P.J., Skirving, W., Strong, A.E., Hardy, J.T., LeDrew, E.F., Hochberg, E.J., Stumpf, R.P. and David, L.T. (2004). Remote sensing of coral reefs and their physical environment. Mar. Poll. Bull. 48: 219-228.

NASA (2010). Research News. 2009: Second Warmest Year on Records; End of Warmest Decade. http://www.giss.nasa.gov/research/news/20100121/.

NASA (2013). GISS Surface Temperature analysis (GISTEMP). http://data.giss.nasa.gov/ gistemp/.

Nativi, S., Craglia, M. and Pearlman, J. (2012). The brokering approach for multidisciplinary interoperability: A position paper, Int. J. Spatial Data Infrastructures Res. 7:1–15.

Nativi, S., Craglia, M. and Pearlman, J. (2013). Earth science infrastructures interoperability: The brokering approach, IEEE JSTARS, Vol. 6 N. 3, pp. 1118-1129.

Nativi, S., Domenico, B., Caron, J., Davis, E. and Bigagli, L. (2006). Extending THREDDS middleware to serve OGC community. Adv. Geosci. 8: 57-62.

Nativi, S., Khalsa, S.J., Domenico, B., Craglia, M., Pearlman, J., Mazzetti, P. and Rew, R. (2011). The Brokering Approach for Earth Science Cyberinfrastructure, EarthCube White Paper. U.S. NSF [Online]. http://semanticcommunity.info/@api/deki/files/13798/=010_Domenico.pdf.

Nativi, S., Mazzetti, P. and Geller, G.N. (2012). Environmental model access and interoperability: The GEO Model Web initiative, Environ. Modell. Softw. 39: 214–228.

Nerem, R.S., Chambers, D.P., Choe, C. and Mitchum, G.T. (2010). Estimating mean sea level change from the TOPEX and Jason altimeter missions. Mar. Geod. 33(S1): 435-446, 10.1080/01490419. 2010.491031.

Nobre, P., De Almeida, R.A.F., Malagutti, M. and Giarolla, E. (2012). Coupled ocean-atmosphere variations over the South Atlantic Ocean. J. Climate 25: 6349-6358.

Nobre, P., Siqueira, L.S.P., de Almeida, R.A.F., Malagutti, M., Giarolla, E., Castelão, G.P., Bottino, M.J., Kubota, P., Figueroa, S.N., Costa, M.C., Baptista, M. Jr., Irber, L. Jr. and Marcondes, G.G. (2013). Climate simulation and change in the Brazilian Climate Model. J. Climate 26: 6716–6732.

Nonaka, I. and Takeuchi, H. (1995). The knowledge creating company: how Japanese companies create the dynamics of innovation. New York: Oxford University Press, Inc.

NSDIC (2013a). Sea Ice Index. http://nsidc.org/data/docs/noaa/g02135 _seaice_index/ index.html#processingstep.

NSDIC (2013b). State of the Cryosphere. Sea ice. http://nsidc.org/ cryosphere/sotc/sea_ice. html.

NSDIC (2013c). State of the Cryosphere. Ice Shelves. http://nsidc.org/ cryosphere/sotc/ iceshelves.html.

O'Neill, J.M., Davis, T.W., Burford, M.A., and Gobler, C.J. (2012). The rise of harmful cyanobacteria blooms: the potential roles of eutrophication and climate change. Harmful Algae 14: 313-334.

Oddo, P. and Leth, O.K. (2012). Coastal and Shelf Seas Modeling Working Group, EuroGOOS 19th Annual General Meeting, http:// www.eurogoos.org/documents/eurogoos/downloads/eg12_39coastal_m odellingwg.pdf.

Oke, P.R., Brassington, G.B., Griffin, D.A. and Schiller, A. (2008). The Bluelink ocean data assimilation system (BODAS). Oc. Modelling 01/2008, 10.1016/j.ocemod.2007.11.002.

Olson, R.J., Shalapyonok, A.A. and Sosik, H.M. (2003). An automated submersible flow cytometer for analyzing pico- and nanophytoplankton: FlowCytobot. Deep-Sea Res. Pt I 50: 301-315.

Ono, T., Saino, T., Kurita, N. and Sasaki, K. (2004). Basin-scale extrapolation of shipboard pCO_2 data by using satellite SST and Chla. Int. J. Remote Sens. 25(19): 3803–3815, 10.1080/0143116031000165 7515.

Öström, B. (1976). Fertilization of the Baltic by nitrogen fixation in the blue-green alga *Nodularia spumigena*. Remote Sens. Environ. 4: 305–310.

Oubelkheir, K., Clementson, L., Webster, I., Ford, P., Dekker, A.G., Radke, L. and Daniel, P. (2006). Using inherent optical properties to investigate biogeochemical dynamics in a tropical macrotidal coastal system. J. Geophys. Res. 111, C07021, 10.1029/2005JC003113.

Paerl, H., Yin, K. and Cloern, J. (2011). Global patterns of phytoplankton dynamics in coastal ecosystems. Eos, Trans. Amer. Geophys. Union 92(10), 85. doi:10.1029/2011EO100007.

Pandolfi, J.M., Bradbury, R.H., Sala, E., Hughes, T., Bjorndal K.A., Cooke, R., McArdle, D., Newman, M.L.M., Paredes, G., Warner, R. and Jackson, J. (2003). Global trajectories of the long-term decline of coral reef ecosystems. Science 301: 955-958.

Park, G.-H., Wanninkhof, R., Doney, S.C., Takahashi, T., Lee, K., Feely, R.A., Sabine, C.L., Triñanes, J. and Lima, I.D. (2010). Variability of global net sea-air CO_2 fluxes over the last three decades using empirical relationships. Tellus B 62(5): 352–368, doi:10.1111/j.1600-0889.2010.00498.x.

Parslow, J., Cressie, N., Campbell, E.P., Jones, E. and Murray, L. (2013). Bayesian learning and predictability in a stochastic nonlinear dynamical model. Ecological Applications 23(4): 679-698.

Pauly, D. and Christensen, V. (1993). Graphical representation of steady-state trophic ecosystem models, p. 20-28. In V. Christensen and D. Pauly (eds.) Trophic models of aquatic ecosystems. ICLARM Conf. Proc. 26, 390 p.

Pauly, D. and Christensen, V. (1995). Primary production required to sustain global fisheries. Nature 374: 255-257.

Peltier, W.R. (2009). Closure of the budget of global sea level rise over the GRACE era: The importance and magnitudes of the required corrections for global isostatic adjustment. Quart. Sci. Rev. 28: 1658-1674, 10.1016/j.quascirev.2009.04.004.

Pendleton, D.E., Pershing, A.J., Brown, M.W., Mayo, C.A., Kenney, R.D., Record, N.R. and Cole, T.V.N. (2009). Regional-scale mean copepod concentration indicates relative abundance of North Atlantic right whales. Mar. Ecol. Prog. Ser. 378: 211-225.

Penven, P., Marchesiello, P., Debreu, L. and Lefevre, J. (2007). Software tools for pre- and post-processing of oceanic regional simulations. Environ. Model. Softw. 23: 660-662.

Pérez, O.M., Ross, L.G., Telfer, T.C. and del Campo Barquin, L.M. (2003). Water quality requirements for marine fish cage site selection in Tenerife (Canary Islands): predictive modelling and analysis using GIS. Aquaculture 224:51-69.

Pershing, A.J., Record, N.R., Monger, B.C., Mayo, C.A., Brown, M.W., Cole, T.V.N., Kenney, R.D., Pendleton, D.E. and Woodard, L.A.

(2009a). Model-based estimates of right whale habitat use in the Gulf of Maine. Mar. Ecol. Prog. Ser. 378: 245-257.

Pershing, A.J., Record, N.R., Monger, B.C., Pendleton, D.E. and Woodard, L.A. (2009b). Model-based estimates of *Calanus finmarchicus* abundance in the Gulf of Maine. Mar. Ecol. Prog. Ser. 378: 227-243.

Pfeffer, W.T. (2011). Land ice and sea level rise: A thirty-year perspective. Oceanography 24: 94-111, 10.5670/oceanog.2011.30.

PICO-I, (2012). Requirements for global implementation of the Strategic Plan for Coastal GOOS. GOOS Report #193, available at http://www.ioc-goos.org/index.php?option=com_oe&task=viewDocumentRecord&docID=7702.

Plag, H.-P. and DiGiacomo P., (2014). The Global Coastal Zone Information System: Supporting the co-design and co-creation of practice-relevant knowledge for deliberative coastal zone governance. In preparation.

Plag, H.-P., Adegoke, J., Bruno, M., Christian, R., DiGiacomo, P., McManus, L., Nicholls, R. and van de Wal, v., (2010). Observations as Decision Support for Coastal Management in Response to Local Sea Level Changes. In Hall, J., Harrison, D.E. & Stammer, D. (eds.): Proceedings of OceanObs'09: Sustained Ocean Observations and Information for Society (Vol. 2), Venice, Italy, 21-25 September 2009, ESA Publication WPP-306, doi:10.5270/OceanObs09.cwp.69.

Plag, H.-P., Foley, G., Jules-Plag, S., Kaufman, J., Ondich, G., (2012). The GEOSS User Requirement Registry: Linking users of GEOSS across disciplines and societal areas. Preliminary Proceed. of IGARSS 2012, Munich, Germany, July 23-27, 2012.

Platt, T., Fuentes-Yaco, C. and Frank, K.T. (2003). Spring algal bloom and larval fish survival. Nature 423: 398-399.

Platt, T., Harrison, W.G., Lewis, M.R., Li, W.K.W., Sathyendranath, S., Smith, R.E. and Vezina, A.F. (1989). Biological production of the oceans: The case for a consensus. Mar. Ecol. Prog. Ser. 52: 77–88.

Platt, T., Sathyendranath, S. and Fuentes-Yaco, C. (2007). Biological oceanography and fisheries management: perspective after 10 years. ICES J. Mar. Sci. 64: 863–869.

Platt, T., Sathyendranath, S., Forget, M.-H., White, G.N., Caverhill, C., Bouman, H., Devred, E. and Son, S. (2008). Operational estimation of primary production at large geographical scales. Remote Sens. Environ. 112(8): 3437–3448, 10.1016/j.rse.2007.11.018.

Polovina, J.J., Balazs, G.H., Howell, E.A., Parker, D.M., Seki, M.P. and Dutton, P.H. (2004). Forage and migration habitat of loggerhead (*Caretta caretta*) and olive ridley (*Lepidochelys olivacea*) sea turtles in the central North Pacific Ocean. Fish. Oceanogr. 13: 36-51.

Polovina, J. J., Howell, E., Kobayashi, D.R. and Seki, M.P. (2001). The transition zone chlorophyll front, a dynamic global feature defined migration and forage habitat for marine resources. Prog. Oceanogr. 49: 469-483.

Polovina, J.J., Howell, E.A. and Abecassis, M. (2008). Ocean's least productive waters are expanding. Geophys. Res. Lett. 35, L03618, 10.1029/2007GL031745.

Polovina, J.J., Kobayashi, D.R., Parker, D.R., Seki, M.P., and Balazs, G.H. (2000). Turtles on the edge: movement of loggerhead turtles (*Caretta caretta*) along oceanic fronts spanning longline fishing grounds in the central North Pacific, 1997-1998. Fish. Oceanogr. 9: 71-82.

Polovina, J.J., Mitchum, G.T., Graham, N.E., Craig, M.P., DeMartini, E.E. and Flint, E.N. (1994). Physical and biological consequences of a climate event in the central North Pacific. Fish. Oceanogr. 3: 15-21.

Polovina, J.J., Uchida, I., Balazs, G., Howell, E.A., Parker, D. and Dutton, P. (2006). The Kuroshio Extension Bifurcation Region: a pelagic hotspot for juvenile loggerhead sea turtles. Deep-Sea Res. Pt II 53: 326-339.

Powell., B.S. and Moore, A.M. (2009). Estimating the 4DVAR analysis error of GODAE products. Ocean Dynamics 59(1): 121-138.

Pretty, J.N., Mason, C.F., Nedwell, D.B., Hine, R.E., Leaf S. and Dils R. (2003). Environmental costs of freshwater eutrophication in England and Wales. Environ. Sci. Technol., 37: 201 –208.

Qin, Y., Brando, V.E., Dekker, A.G. and Blondeau-Patissier, D. (2007). Validity of SeaDAS Water Constituents Retrieval Algorithms in Australian Tropical Coastal Waters. Geophys. Res. Lett. 34, l21603, 10.1029/2007GL030599.

Raine, R., McDermott, G., Silke, J., Lyons, K., Nolan, G. and Cusack, C. (2010). A simple short range model for the prediction of harmful algal events in the bays of southwestern Ireland. J. Marine Syst. 83: 150-157.

Reul, N., Saux-Picart, S., Chapron, B., Vandemark, D., Tournadre, J. and Salisbury, J. (2009). Demonstration of ocean surface salinity microwave measurements from space using AMSR-E data over the Amazon plume. Geophys. Res. Lett. 36, 10.1029/2009GL038860.

Rigor I.G., Wallace, J.M. and Colony, L. (2002). Response of Sea Ice to the Arctic Oscillation. J. Climate 15: 2648-2663, 10.1175/1520-0442 (2002)015.

Rigor, I.G. and Wallace, J.M. (2004). Variations in the age of Arctic sea-ice and summer sea-ice extent. Geophys. Res. Lett. 31, L09401, 10.1029/2004GL019492.

Rind, D., Chandler, M., Lerner, J., Martinson, D.G. and Yuan, X. (2001). Climate response to basin-specific changes in latitudinal temperature gradients and implications for sea ice variability. J. Geophys. Res. 106, D17, 20161-20173, 10.1029/2000JD900643.

Roemmich, D. and Cornuelle, B. (1992). The subtropical mode waters of the South Pacific Ocean, J. Phys. Ocean., 22: 1178-1187.

Roemmich, D., Gilson, J., Willis, J., Sutton, P. and Ridgway, K. (2005). Closing the time-varying mass and heat budgets for large ocean areas: The Tasman Box. J. Climate 18 (13): 2330–2343.

Santamaría-del-Angel, E., Kampel, M., Lutz, V., Millán-Núñez, R., González-Silvera, A., Dogliotti, A.I., Frouin, R., Muller-Karger, F., and all Antares Network. (2010). Climate change evaluated at marine time-series stations. The Antares Network an effort of the Americas in long term studies. Conferencia 'American Geophysical Union (AGU)'. Foz do Iguazu, Brasil. Agosto de 2010.

Santoro, M., Mazzetti, P., Nativi, S., Fugazza, C., Granell, C. and Diaz, L. (2012). Methodologies for augmented discovery of geospatial resources. In: Discovery of Geospatial Resources: Methodologies, Technologies, and Emergent Applications, IGI Global, 335 pp. Web. 30 Mar. 2012. doi:10.4018/978-1-4666-0945-7.

Sarma, V.V.S.S., Saino, T., Sasaoka, K., Nojiri, Y., Ono, T., Ishii, M., Inoue, H.Y. and Matsumoto, K. (2006). Basin-scale pCO_2 distribution using satellite sea surface temperature, Chla, and climatological salinity in the North Pacific in spring and summer. Global Biogeochem. Cy. 20(3), 10.1029/2005GB002594.

Sathyendranath, S., Ahanhanzo, J., Bernard, S., Byfield, V., Dowell, M., Field, J., Groom, S., Hardman-Mountford, N., Hoepffner, N., Jacobs, T., Kampe, M., Kumar, S., Lutz, V. and Platt T. (2010). ChloroGIN: Use of satellite and *in situ* data in support of ecosystem-based management of marine ecosystems. In: Proceedings of OceanObs'09: Sustained Ocean Observations and Information for Society (Vol. 2), Hall, J., Harrison, D.E. and Stammer, D. (eds.), Venice, Italy, 21-25 September 2009, ESA Publication WPP-306.

Scambos, T.A., Bohlander, J.A., Shuman, C.A. and Skvarca, P. (2004). Glacier acceleration and thinning after ice shelf collapse in the Larsen

B embayment, Antarctica. Geophys. Res. Lett. 31, L18402, 10.1029/ 2004GL020670.

Scambos, T.A., Fricker, H.A., Liu, C.-C., Bohlander, J., Fastook, J., Sargent, A., Massom, A. and Wu, A.-M. (2009). Ice shelf disintegration by plate bending and hydro-fracture: Satellite observations and model results of the 2008 Wilkins ice shelf break-ups. Earth Planet. Sci. Lett. 280: 51-60, 10.1016/j.epsl.2008.12. 027.

Scambos, T.A., Haran, T.M., Fahnestock, M.A., Painter, T.H. and Bohlander, J.A. (2007). MODIS-based Mosaic of Antarctica (MOA) data sets: Continent-wide surface morphology and snow grain size. Remote Sens. Environ. 111: 242-257, 10.1016/j.rse.2006.12.020.

Scatasta, S., Stolte, W., Graneli, E., Weikard, H.P. and Van Ierland, E. (2003). The socio-economic impact of harmful algal blooms in European Union countries. Fifth deliverable for the project ECOHARM Contract No. EVK3-CT-2001-80003.

Schiemeier, Q. (2013). Oceans under surveillance: Three projects seek to track changes in the Atlantic overturning circulation currents. Nature 497: 167-168, 10.1038/497167a.

Schiller, A. and Brassington, G.B. (eds.) (2011). Operational Oceanography in the 21st Century, DOI 10.1007/978-94-007-0332-2_1, Springer Science+Business Media B.V., http://www.springer. com/earth+sciences+and+geography/oceanography/book/978-94-007-0331-5.

Schlüter, M., Kraberg, A. and Wiltshire, K.H. (2012). Long-term changes in the seasonality of selected diatoms related to grazers and environmental conditions. J. Sea Res. 67: 91-97.

Scholin, C.A., Doucette, G.J. and Cembella, A.D. (2008). Prospects for developing automated systems for *in situ* detection of harmful algae and their toxins. In: Real–Time Coastal Observing Systems for Ecosystem Dynamics and Harmful Algal Blooms, Babin, M., Roesler, C.S. and Cullen, J.J. (eds.), UNESCO Publishing, Paris, France, pp. 413–462.

Schroeder, T., Devlin, M.J., Brando, V.E., Dekker, A.G., Brodie, J.E., Clementson, L.A. and McKinna, L. (2012). Inter-annual variability of wet season freshwater plume extent into the Great Barrier Reef lagoon based on satellite coastal ocean colour observations. Mar. Poll. Bull. 65(4-9): 210-223.

Sellner, K.G., Doucette, G.J. and Kirkpatrick, G.J. (2003). Harmful algal blooms: causes, impacts and detection. J. Indian Microbiol. Biotechnol. 30: 383–406.

Serreze, M.C., Holland, M.M. and Stroeve, J. (2007). Perspectives on the Arctic's shrinking sea-ice cover. Science 315: 1533-1536, 10.1128/science.1139426.

Sherman, K., Sissenwine, M., Christensen, V., Duda, A., Hempel, G., Ibe, C., Levin, S., Lluch-Belda, D., Matishov, G., McGlade, J., O'Toole, M., Seitzinger, S., Serra, R., Skjoldal, H.R., Tang, Q., Thulin, J., Vandeweerd, V. and Zwanenburg, K. (2005). A global movement toward an ecosystem aproach to management of marine resources. Mar. Ecol. Prog. Ser. 300: 275-279.

Shulman, I., Kindle, J., Martin, P., deRada, S., Doyle, J., Penta, B., Anderson, S., Chavez, F., Paduan, J. and Ramp, S. (2007). Modeling of upwelling/relaxation events with the Navy Coastal Ocean Model, J. Geophys. Res., 112, C06023, doi:10.1029/2006JC003946.

Silver, M.W., Bargu, S., Coale, S.L., Benitez-Nelson, C.R., Garcia, A.C., Roberts, K.J., Sekula-Wood, E., Bruland, K.W. and Coale, K.H. (2010). Toxic diatoms and domoic acid in natural and iron enriched waters of the oceanic Pacific. P. Natl Acad. Sci. USA, 107(48): 20762-20767.

Slater, T. (2012). What is Interoperability?, Network Centric Operations Industry Consortium – NCOIC, 2012.

Smith, N. and Lefebvre, M. (1997). Developing a strategy for GODAE, GODAE strategy papers, GODAE website, http://www.godae.org/modules/documents/documents/Smith-Lefebvre-1997.pdf).

Sobek, S., Algesten, G., Bergstrom, A.K., Jansson, M. and Tranvik, L.J. (2003). The catchment and climate regulation of pCO_2 in boreal lakes. Glob. Change Biol. 9(4): 630–641, 10.1046/j.1365-2486.2003.00619.x.

Solanki, H.U., Dwivedi, R.M., Nayak, S.R., Somvanshi, V.S., Gulati, D.K. and Pattnayak, S.K. (2003). Fishery forecast using OCM chlorophyll concentration and AVHRR SST: validation results off Gujarat coast, India. Int. J. Remote Sens. 24: 3691-3699.

Soltwedel, T., Bauerfeind, E., Bergmann, M., Budaeva, N., Hoste, E., Jaeckisch, N., Mathiessen, J., Mokiewsky, V., Nöthig, E.-M., Quéric, N.-V., Sablotny, B., Sauter, E., Schewe, I., Urban-alinga, B., Wegner, J., Wlodarska-Kowalczuk, M., and Klages, M. (2005). Hausgarten: Multidisciplinary investigations at a deep-sea, long-term observatory in the Arctic Ocean. Oceanography 18: 47-61.

Son, Y.B., Min, J.E. and Ryu, J.H. (2012). Detecting Massive Green Algae (*Ulva prolifera*) Blooms in the Yellow Sea and East China Sea using Geostationary Ocean Color Imager (GOCI). Data. Ocean Sci. J. 47(3): 359-375.

Steffen, K., Thomas, R., Rignot, E., Cogley, G., Dyurgerov, M., Raper, S., Huybrechts, P. and Hanna, E. (2010). Cryospheric contributions to sea level variability. In: Understanding sea-level rise and variability, Church, J.A., Woodworth, P.L., Aarup, T. and Wilson, W.S. (eds.), Wiley-Blackwell Publishing Limited., pp. 177-225.

Steinberg, C.E.W., Hoss, S., Kloas, W., Lutz, I., Meinelt, T., Pflugmacher, S. and Wiegand, C. (2004). Hormonelike effects of humic substances on fish, amphibians, and invertebrates. Environ. Toxicol. 19: 409–411.

Stewart, R.R. (2005). Our Ocean Planet – oceanography in the 21st century, http://oceanworld.tamu.edu/resources/oceanography-book/ carboncycle.htm.

Strong, A.E., Arzayus, F., Skirving, W. and Heron, S.F. (2006). Identifying coral bleaching remotely via Coral Reef Watch: Improving integration and implications for changing climate. In: Phinney, J.T. (ed), Coral Reefs and Climate Change: Science Management, AGU, Washington, DC, pp. 163-180.

Stuart, V., Platt, T., and Sathyendranath, S. (2011). The future of fisheries science in management: a remote-sensing perspective. – ICES Journal of Marine Science, doi:10.1093/icesjms/fsq200.

Stumpf, R.P., Litaker, R.W., Lanerolle, L. and Tester, P.A. (2008). Hydrodynamic accumulation of *Karenia* off the west coast of Florida. Cont. Shelf Res. 28: 189–213.

Stumpf, R.P. and Tomlinson, M.C. (2005). Remote sensing of harmful algal blooms. In: Remote Sensing of Coastal Aquatic Environments, Miller, R.L., Del Castillo, C.E. and Mckee, B.A., Springer Netherlands, pp. 277-296.

Sutton, P. and Roemmich, D. (2001). Ocean temperature climate off north-east New Zealand, New Zeal. J. Mar. Fresh. 35: 553-565.

Sverdrup, H.U. (1953). On conditions for the vernal blooming of phytoplankton. J. Conseil, 18: 287–295.

Swart, S., Speich, S., Ansorge, I.J., Goni, G.J., Gladyshev, S. and Lutjeharms, J.R.E. (2008). Transport and variability of the Antarctic Circumpolar Current south of Africa, J. Geophys. Res., 113, C09014, doi:10.1029/2007JC004223.

Szulanski, G. (1996). Exploring internal stickiness: Impediments to the transfer of best practice within the firm. Strategic Manage. J. 17(1): 27-44.

Takeoka, H. (2002). Progress in Seto Inland Sea Research. J. Oceanogr. 58: 93-107.

Taylor, J. and Ferrari, R. (2011). A shutdown of turbulent convection can trigger the spring phytoplankton bloom. Limnol. Oceanogr. 56: 2293-2307.

Tranvik, L.J (1990). Bacterioplankton growth on fractions of dissolved organic carbon of different molecular weights from humic and clear waters. Appl. Environ. Microbiol. 56: 1672-1677.

Tranvik, L.J., Downing, J.A., Cotner, J.B., Loiselle, S.A., Striegl, R.G., Ballatore, T.J., Dillon, P., Finlay, K., Fortino, K., Knoll, L.B., Kortelainen, P.L., Kutser, T., Larsen, S., Laurion, I., Leech, D.M., McCallister, S.L., McKnight, D.M., Melack, J.M., Overholt, E., Porter, J.A., Prairie, Y., Renwick, W.H., Roland, F., Sherman, B.S., Schindler, D.W., Sobek, S., Tremblay, A., Vann,i M.J., Verschoor, A.M., von Wachenfeldt, E., and Weyhenmeyer G.A. (2009). Lakes and reservoirs as regulators of carbon cycling and climate. Limnol. Oceanogr. 54: 2298-2314.

Trenberth, K.E., Jones, P.D., Ambenje, P.G., Bojariu, R., Easterling, D.R., Klein Tank, A.M.G., Parker, D.E., Renwick, J.A., Rusticucci, M., Soden, B. and Zhai, P. (2007). Surface and atmospheric climate change. Climate Change 2007: The Physical Science Basis. Contribution of Working Group I to the Fourth Assessment Report of the Intergovernmental Panel on Climate Change, Cambridge University Press, Cambridge 235-336.

Tsubouchi, T., Suga, T. and Hanawa, K. (2007). Three types of South Pacific Subtropical Mode Waters: Their relation to the large-scale circulation of the South Pacific subtropical Gyre and their temporal variability. J. Phys. Oceanogr. 37: 2478-2490.

UNEP (2005). Pacific island villagers first climate change refugees. UNEP News Release. United Nations Environment Programme, Nairobi. http://www.grida.no/news/press/1533.aspx.

Vaccari, L., Craglia, M., Fugazza, C., Nativi, S. and Santoro, M. (2012). Integrative research: The EuroGEOSS experience, IEEE J. Sel. Topics Appl. Earth Observ. Remote Sens. 5(6): 1603–1611.

Vahtera, E., Comnley, D.J., Gustafsson, B.G., Kuosa, H., Pitkänen, H., Savchuk, O.P., Tamminen, T., Viitasalo, M., Voss, M., Wasmund, N. and Wulff, F. (2007). Internal ecosystem feedbacks enhance nitrogen-fixing cyanobacteria blooms and complicate management in the Baltic Sea. Ambio 36: 186–194.

Vandepitte, L., Vanhoorne, B., Kraberg, A., Anisimova, N., Antoniadou, C., Araujo, R., Bartsch, I., Beker, B., Benedetti-Cechi, L., Bertocci, I.,

Cochrane, S., Cooper, K., Craeyemeersch, J., Crisp, D.J., Cristou, E., Dahle, S., de Boissier, M., de Kluijver, M., Denisenko, S., De Vito, D., Duineveld, G., Escaravage, V., Fleischer, D., Fraschetti, S., Giangrande, A., Heip, C., Hummel, H., Janas, U., Karez, R., Kedra, M., Kingston, P., Kuhlenkamp, R., Libes, M., Martens, P., Mieszkowska, N., Mudrak, S., Munda, I., Orfanidis, S., Orlando-Bonaca, M., Palerud, R., Rachor, E., Reichert, K., Rumohr, H., Schiedek, D., Schubert, P., Sistermans, W.C.H., Sousa Pinto, I., Southward, A.J., Terlizzi, A., Tsiaga, E., van Beusekom, J.E.E., Warzocha, J., Wasmund, N., Weskawski, JM., Widdicombe, C., Wlodarska, K., Vanden Berghe, E., and Zettler, M. (2010). Data integration for European marine biodiversity research: creating a database on benthos and plankton to study large-scale patterns and long-term changes. Hydrobiologia 644: 1-13.

Velo-Suarez, L. and Gutierrez-Estrada, J.C. (2007). Artificial neural network approaches to one-step weekly prediction of *Dinophysis acuminata* blooms in Huelva (Western Andalucıa, Spain). Harmful Algae 6(3): 361-671.

Velo-Suarez, L., Reguera, B., Gonzalez-Gil, S., Lunven, M., Lazure, P., Nezan, E. and Gentien, P. (2010). Application of a 3D Lagrangian model to explain the decline of a *Dinophysis acuminata* bloom in the Bay of Biscay. J. Marine Syst. 83: 193-202.

Wakelin, S.L., Holt, J.T. and Proctor, R. (2009). The influence of initial conditions and open boundary conditions on shelf circulation in a 3D ocean-shelf model of the North East Atlantic. Ocean Dynamics 59(1): 67-81.

Watson, A.J., Schuster, U., Bakker, D.C.E., Bates, N.R., Corbiere, A., Gonzalez-Davila, M., Friedrich, T., Hauck, J., Heinze, C., Johannessen, T., Körtzinger, A., Metzl, N., Olafsson, J., Olsen, A., Oschlies, A., Padin, X.A., Pfeil, B., Santana-Casiano, J.M., Steinhoff, T., Telszewski, M., Rios, A.F., Wallace, D.W.R. and Wanninkhof, R. (2009). Tracking the variable North Atlantic sink for atmospheric CO_2. Science 326(5958), 1391–1393, 10.1126/science.1177394.

Wetzel, R.G. (2001). Limnology: lake and river ecosystems. Elsevier, 3rd Edition, 1006 pp.

Whaling Commission (2001). Report of the workshop on the comprehensive assessment of right whales: A worldwide comparison, Best, P.B., Bannister, J.L., Brownell, R.L. Jr. and Donovan, G.P. (eds), J. Cet. Res. Manage. Special Issue (2): 1-35.

Widdicombe, C.E., Eloire, D., Harbour D., Harris R.P. and Somerfield, P.J. (2010). Long-term phytoplankton community dynamics in the Western English Channel. J. Plankton Res. 32: 643-655.

Wijffels, S. and Meyers, G. (2004). An intersection of oceanic waveguides: variability in the Indonesian Throughflow region, J. Phys. Oceanogr. 34: 1232-1254.

Willis, J.K. and Church, J.A. (2012). Regional Sea Level Projection. Science 336: 550-551, 10.1126/science.1220366.

Wilson, C. and Adamec, D. (2001). Correlations between surface chlorophyll and sea surface height in the tropical Pacific during the 1997-1999 El Niño-Southern Oscillation event. J. Geophys. Res. 106: 31175-31188.

Wiltshire, K.H., Kraberg, A., Bartsch, I., Boersma, M., Franke, H.D., Freund, J., Gebühr, C., Gerdts, G., Stockmann, K. and Wichels, A. (2010). Helgoland Roads: 45 years of change in the North Sea. Estuaries Coasts 33: 295-310.

Wrigley, R.C. and Horne, A.J. (1974). Remote sensing and lake eutrophication. Nature 250: 213–214.

Yeh, S.-W., Kug, J.-S, Dewitte, B., Kwon, M.-H., Kirtman, B.P. and Jin, F.-F. (2009). El Niño in a changing climate. Nature 461: 512-515, 10.1038/nature08316.

Zainuddin, M., Saitoh, S.-i. and Saitoh, K. (2004). Detection of potential fishing ground for albacore tuna using synoptic measurements of ocean color and thermal remote sensing in the northwestern North Pacific. Geophys. Res. Lett. 31: L20311, doi: 20310.21029/22004GL021000.

Zhai, L., Gudmundsson, K., Miller, P., Peng, W.J., Gudfinnsson, H., Debes, H., Hatun, H., White, G.N. III, Hernandez Walls, R., Sathyendranath, S. and Platt, T. (2012). Phytoplankton phenology and production around Iceland and Faroes. Cont. Shelf Res., doi:10.1016/j.csr.2012.01.013.

Zhai, L., Platt, T., Tang, C., Sathyendranath, S. and Hernández Walls, R. (2011). Phytoplankton phenology on the Scotian Shelf. ICES Journal of Marine Science. doi:10.1093/icesjms/fsq175.

Zhai, L., Platt, T., Tang, C., Sathyendranath, S. and Walne, A. (2013). The response of phytoplankton to climate variability associated with the North Atlantic Oscillation. Deep-Sea Res. II, http://dx.doi.org/10.1016/j.dsr2.2013.04.009.

ANNEX IV

LIST OF AUTHORS AND AFFILIATIONS

Name	Surname	Affiliation
José C.	Báez	Instituto Español de Oceanografía (IEO), Malaga, España
Molly	Baringer	Atlantic Oceanographic and Meteorological Laboratory, National Oceanic and Atmospheric Administration (NOAA), United States
Mike	Bell	Meteorological Office, Exeter, United Kingdom
Stewart	Bernard	NRE Earth Observation, Council for Scientific and Industrial Research (CSIR) and Department of Oceanography, University of Cape Town, South Africa
Vittorio	Brando	Land and Water, Commonwealth Scientific and Industrial Research Organisation (CSIRO), Brisbane, Australia
Edmo	Campos	Instituto Oceanográfico, Universidade de São Paulo (USP), Brazil
Prakash	Chauhan	Space Application Centre, Indian Space Research Organisation (ISRO), Gujarat, India
Victoria	Cheung	Partnership for Observation of the Global Oceans, Plymouth, United Kingdom
Lesley	Clementson	Land and Water, Commonwealth Scientific and Industrial Research Organisation (CSIRO), Brisbane, Australia
Rebecca	Cowley	Marine and Atmospheric Research, Commonwealth Scientific and Industrial Research Organisation (CSIRO), Tasmania, Australia
Massimo	Craglia	European Commission (EC), Joint Research Centre (JRC), Institute for Environment and Sustainability, Italy
Douglas	Cripe	Group on Earth Observations Secretariat (GEO), Geneva, Switzerland
Pierre	de Mey	Centre National de la Recherche Scientifique (CNRS), Laboratoire d'Etudes en Géophysique et Océanographie Spatiales (LEGOS), Toulouse, France

Arnold	Dekker	Land and Water, Commonwealth Scientific and Industrial Research Organisation (CSIRO), Brisbane, Australia
Paul	DiGiacomo	Satellite Oceanography and Climatology Division, Center for Satellite Applications and Research (STAR), National Oceanic and Atmospheric Administration (NOAA), United States
Samy	Djavidnia	EU European Maritime Safety Agency (EMSA), Lisbon, Portugal
Eric	Dombrowski	Mercator Ocean, France
Mark D.	Dowell	European Commission (EC), Joint Research Centre (JRC), Institute for Environment and Sustainability, Italy
Carlos	Garcia Soto	Instituto Español de Oceanografia (IEO), Madrid, España
Gustavo	Goni	Atlantic Oceanographic and Meteorological Laboratory, National Oceanic and Atmospheric Administration (NOAA), United States
Watson W.	Gregg	Goddard Space Flight Center (GSFC), National Aeronautics Space Administration (NASA), United States
Laura	Griesbauer	United States Integrated Ocean Observing System, National Oceanic and Atmospheric Administration (NOAA), United States
Ann	Gronell Thresher	Marine and Atmospheric Research, Commonwealth Scientific and Industrial Research Organisation (CSIRO), Tasmania, Australia
Steve	Groom	Plymouth Marine Laboratory, United Kingdom
Nick	Hardman-Mountford	Marine and Atmospheric Research, Commonwealth Scientific and Industrial Research Organisation (CSIRO), Perth, Australia
Jonathan	Hodge	Land and Water, Commonwealth Scientific and Industrial Research Organisation (CSIRO), Brisbane, Australia
Nicolas	Hoepffner	European Commission (EC), Joint Research Centre (JRC), Institute for Environment and Sustainability, Italy
Joji	Ishizaka	Hydrospheric Atmospheric Research Centre (HyARC), University of Nagoya, Nagoya, Japan
Johnny	Johannessen	Nansen Environmental and Remote Sensing Centre, Bergen, Norway

Emyln	Jones	Land and Water, Commonwealth Scientific and Industrial Research Organisation (CSIRO), Brisbane, Australia
Milton	Kampel	Divisão de Sensoriamento Remoto (DSR), Instituto Nacional de Pesquisas Espaciais (INPE), São José dos Campos, Brazil
Edward	King	Land and Water, Commonwealth Scientific and Industrial Research Organisation (CSIRO), Brisbane, Australia
Villy	Kourafalou	Rosenstiel School of Marine and Atmospheric Science, University of Miami, United States
Alexandra	Kraberg	Alfred Wegner Institute for Polar and Marine Research, Helgoland, Germany
Raphael	Kudela	Ocean Science Department, University of California Santa Cruz, United States
Tiit	Kutser	Estonian Marine Institute, University of Tartu, Estonia
Vivian	Lutz	Consejo Nacional de Investigaciones Científicas y Técnicas (CONICET), Instituto Nacional de Investigación y Desarrollo Pesquero (INIDEP), Argentina
Jose	Muelbert	Instituto de Oceanografia, Universidade Federal do Rio Grande (FURG), Rio Grande, Brazil
Stefano	Nativi	Istituto sull'Inquinamento Atmosferico (IIA), Consiglio Nazionale delle Richerche (CNR), Italy
Paulo	Nobre	Centro de Previsão de Tempo e Estudos Climáticos, Instituto Nacional de Pesquisas Espaciais (INPE), Brazil
Hans-Peter	Plag	Climate Change and Sea Level Rise Initiative (CCSLRI), Old Dominion University, United States
Trevor	Platt	Partnership for Observation of the Global Oceans, PML, Plymouth (POGO), United Kingdom
Jeffrey	Polovina	Pacific Islands Fisheries Science Center (PIFSC), National Marine Fisheries Service (NMFS), National Oceanic and Atmospheric Administration (NOAA), United States
Dean	Roemmich	Scripps Institution of Oceanography, University of California San Diego, United States
Barbara	Ryan	Group on Earth Observations Secretariat (GEO), Geneva, Switzerland
Joo-Hyung	Ryu	Korea Ocean Research and Development Institute (KORDI), Korea
Shubha	Sathyendranath	Plymouth Marine Laboratory, United Kingdom

Kerry	Sawyer	Committee on Earth Observation Satellites (CEOS), National Oceanic and Atmospheric Administration (NOAA), United States
Angela	Schäfer	Alfred Wegner Institute for Polar and Marine Research, Bremerhaven Germany
Andreas	Schiller	Marine and Atmospheric Research, Commonwealth Scientific and Industrial Research Organisation (CSIRO), Tasmania, Australia
Thomas	Schroeder	Land and Water, Commonwealth Scientific and Industrial Research Organisation (CSIRO), Brisbane, Australia
Sophie	Seeyave	Partnership for Observation of the Global Oceans (POGO), PML, Plymouth, United Kingdom
Janet	Sprintall	Scripps Institution of Oceanography, University of California San Diego, United States
Andy	Steven	Land and Water, Commonwealth Scientific and Industrial Research Organisation (CSIRO), Brisbane, Australia
Venetia	Stuart	International Ocean Colour Coordinating Group (IOCCG) Project Office, Canada
Clemente A. S.	Tanajura	Instituto de Física, Universidad Federal de Bahia (UFBA), Brazil and Ocean Science Department, University of California Santa Cruz, United States
Ariel H.	Troisi	Servicio de Hidrografía Naval, Buenos Aires, Argentina and UNESCO/IOC International Oceanographic Data and information Exchange (IODE).
Lourdes	Velo Suarez	IFREMER, Brest, France and Applied Ocean Physics and Engineering Department, Woods Hole Oceanographic Institution (WHOI), United States
Augustus	Vogel	Office of Naval Research Global (ONRG), United States
Diane E.	Wickland	Earth Science Division, Science Mission Directorate, National Aeronautics Space Administration (NASA), United States
Zdenka	Willis	United States Integrated Ocean Observing System, National Oceanic and Atmospheric Administration, United States
Kirsten	Wilmer Becker	Meteorological Office, Exeter, United Kingdom

Cara	Wilson	Southwest Fisheries Science Center (SWFSC), National Marine Fisheries Service (NMFS), National Oceanic and Atmospheric Administration (NOAA), United States
Li	Zhai	Bedford Institute of Oceanography, Nova Scotia, Canada